FATIGUE DESIGN AND RELIABILITY

Other titles in the ESIS Series

EGF 1	*The Behaviour of Short Fatigue Cracks* Edited by K. J. Miller and E. R. de los Rios
EFG 2	*The Fracture Mechanics of Welds* Edited by J. G. Blauel and K.-H. Schwalbe
EGF 3	*Biaxial and Multiaxial Fatigue* Edited by M. W. Brown and K. J. Miller
EGF 4	*The Assessment of Cracked Components by Fracture Mechanics* Edited by L. H. Larsson
EGF 5	*Yielding, Damage, and Failure of Anisotropic Solids* Edited By J. P. Boehler
EGF 6	*High Temperature Fracture Mechanisms and Mechanics* Edited by P. Bensussan and J. P. Mascarell
EGF 7	*Environment Assisted Fatigue* Edited by P. Scott and R. A. Cottis
EGF/ESIS 8	*Fracture Mechanics Verification by Large Scale Testing* Edited by K. Kussmaul
ESIS/EFG 9	*Defect Assessment in Components – Fundamentals and Applications* Edited by J. G. Blauel and K.-H. Schwalbe
ESIS 10	*Fatigue under Biaxial and Multiaxial Loading* Edited by K. Kussmaul, D. L. McDiarmid, and D. F. Socie
ESIS 11	*Mechanics and Mechanisms of Damage in Composites and Multi-Materials* Edited by D. Baptiste
ESIS 12	*High Temperature Structural Design* Edited by L. H. Larsson
ESIS 13	*Short Fatigue Cracks* Edited by K. J. Miller and E. R. de los Rios
ESIS 14	*Mixed-Mode Fatigue and Fracture* Edited by H. P. Rossmanith and K. J. Miller
ESIS 15	*Behaviour of Defects at High Temperatures* Edited by R. A. Ainsworth and R. P. Skelton
ESIS 16	*Fatigue Design* Edited by J. Solin, G. Marquis, A Siljander, and S. Sipilä
ESIS 17	*Mis-Matching of Welds* K.-H. Schwalbe and M. Koçak
ESIS 18	*Fretting Fatigue* Edited by R. B. Waterhouse and T. C. Lindley
ESIS 19	*Impact and Dynamic Fracture of Polymers and Composites* Edited by J. G. Williams and A. Pavan
ESIS 20	*Evaluating Material Properties by Dynamic Testing* Edited by E. van Walle
ESIS 21	*Multiaxial Fatigue & Design* Edited by A. Pinian, G. Cailletand and T. C. Lindley
ESIS 22	*Fatigue Design of Components* Edited by G. Marquis and J. Solin

For information on how to order titles 1–22, please contact MEP Ltd, Northgate Avenue, Bury St Edmonds, Suffold, IP32 6BW, UK

FATIGUE DESIGN AND RELIABILITY

Editors: G. Marquis and J. Solin

ESIS Publication 23

This volume represents a selection of papers presented at the Third International Symposium on Fatigue Design, FD' 98, held in ESPOO, Finland on 26–29 May, 1998. The meeting was organized by VTT Manufacturing Technology and co-sponsored by the European Structural Integrity Society (ESIS). Partial funding for the event was provided by the European Commission.

1999

Elsevier

Amsterdam • Lausanne • New York • Oxford • Shannon • Singapore • Tokyo

ELSEVIER SCIENCE Ltd
The Boulevard, Langford Lane
Kidlington, Oxford OX5 1GB, UK

First edition 1999

British Library Cataloguing in Publication Data
A catalogue record from the British Library has been applied for.

Library of Congress Cataloging-in-Publication Data

International Symposium on Fatigue Design (3rd : 1998 : Espoo, Finland)
Fatigue design and reliability / editors, G. Marquis and J. Solin.
p. cm. -- (ESIS publication ; 23)
""Selection of papers presented at the Third International Symposium on Fatigue Design, FD '98, held in Espoo, Finland on 26-29 May 1998 ... meeting was organized by VTT Manufacturing Technology and co-sponsored by the European Structural Integrity Society (ESIS)."
ISBN 0-08-043329-4 (hardcover)
1. Materials--Fatigue--Congresses. 2. Structural design--Congresses. 3. Reliability (Engineering)--Congresses.
I. Marquis, G. (Gary) II. Solin, J. III. European Structural Integrity Society. IV. Title. V. Series.
TA418.38.I55 1998
620.1'126--dc21 98-53179
CIP

ISBN: 008 043329 4

∞ The paper used in this publication meets the requirements of ANSI/NISO Z39.48-1992 (Permanence of Paper).
Printed in The Netherlands.

SYMPOSIUM ORGANISERS

Scientific Committee

S. Berge	*Norway*	E. Niemi	*Finland*
J. Bergmann	*Germany*	J. Petit	*France/ESIS*
A. Blom	*Sweden*	J. Polák	*Czech Republic*
M. Brown	*UK*	D. Socie	*USA*
G. Glinka	*Canada*	R. Sunder	*India*
C. Guedes Soares	*Portugal*	K. Wallin	*Finland*
Y Murakami	*Japan*		

International Advisory Board

D. Allen	*European Gas Turbines, UK*
J.-Y. Bérard	*Renault SA, France*
A. Bignonnet	*Peugeot Citroën SA, France*
A. Blarasin	*Fiat Research Center, Italy*
T. Dahle	*ABB Corporate Research, Sweden*
J. Devlukia	*Rover Group, UK*
K.-O. Edel	*Fachhochschule Brandenburg, Germany*
H. Jakubczak	*Warsaw University of Technology, Poland*
B. Johannesson	*Volvo Truck Corp., Sweden*
E. Keim	*Siemens AG, Germany*
V. Köttgen	*LMS Durability Technologies, Germany*
T. Mizoguchi	*Kobe Steel, Japan*
P. Mourilhat	*Electricité de France, France*
V. Panasyuk	*Academy of Sciences, Ukraine*
J. Samuelsson	*VCE Components, Sweden*
C. Sieck	*Caterpillar Inc., USA*
D. Tchankov	*University of Sofia, Bulgaria*
T. Yoshimura	*Toyota Motor Corporation, Japan*

National Advisory Board

H. Hänninen	*Helsinki University of Technology*
E. Pulkkinen	*Ahlström Machinery Corporation*
I. Pusa	*VR Ltd.*
R. Rabb	*Wartsila NSD Ltd.*
K. Rahka	*VTT Manufacturing Technology*
R. Rantala	*Radiation and Nuclear Safety Authority (STUK)*
R. Rintamaa	*VTT Manufacturing Technology*
A. Siljander	*VTT Manufacturing Technology*
S. Sipilä	*Technology Development Centre of Finland*

Local Organisers (VTT Manufacturing Technology)

Gary Marquis, Chairman
Jussi Solin, Co-chairman
Hikka Hänninen
Merja Asikainen
Kari Hyry (TSG-Congress Ltd.)

Elsevier Titles of Related Interest

Books

ABE & TSUTA
AEPA '96: Proceedings of the 3rd Asia-Pacific Symposium on Advances in Engineering Plasticity and its Applications (Hiroshima, August 1996). ISBN 008 042824-X

CARPINTERI
Handbook of Fatigue Crack Propagation in Metallic Structures. ISBN 0-444-81645-3

JONES
Failure Analysis Case Studies. ISBN 008 043338-3

KARIHALOO *ET AL.*
Advances in Fracture Research: Proceedings of the 9th International Conference on Fracture (Sydney, April 1997). ISBN 008 042820-7

KISHIMOTO *ET AL.*
Cyclic Fatigue in Ceramics. ISBN 0-444-82154-6

KLESNIL & LUKÁŠ
Fatigue of Metallic Materials. 2nd Edn. ISBN 0-444-98723-1

LÜTJERING & NOWACK
Fatigue '96: Proceedings of the 6th International Fatigue Congress (Berlin, May 1996). ISBN 008-042268-3

MENCIK
Strength and Fracture of Glass and Ceramics. ISBN 0-444-98685-5

PANASYUK *ET AL.*
Advances in Fracture Resistance and Structural Integrity (ICF 8). ISBN 008-042256-X

RIE & PORTELLA
Low Cycle Fatigue and Elasto-Plastic Behaviour of Materials. ISBN 008 043326-X

Journals

Acta Metallurgica et Materialia
Composite Structures
Computers and Structures
Corrosion Science
Engineering Failure Analysis
Engineering Fracture Mechanics
International Journal of Fatigue
International Journal of Impact Engineering
International Journal of Mechanical Sciences
International Journal of Non-Linear Mechanics
International Journal of Solids and Structures
Journal of Applied Mathematics and Mechanics
Journal of the Mechanics and Physics of Solids
Materials Research Bulletin
Mechanics of Materials
Mechanics Research Communications
NDT&E International
Scripta Metallurgica et Materialia
Theoretical and Applied Fracture Mechanics
Tribology International
Wear
Welding in the World

For more information Elsevier's catalogue can be accessed via the internet on http://www.elsevier.nl

Contents

PREFACE

This volume represents a selection of papers presented at the Third International Symposium on Fatigue Design, Fatigue Design 1998, held in Espoo, Finland on 26-29 May 1998. The meeting was organised by VTT Manufacturing Technology and co-sponsored by the European Structural Integrity Society (ESIS). Partial funding was provided by the European Commission. In attendance were 140 engineers and researchers representing 25 countries.

One objective of the Fatigue Design symposium series has been to help bridge the gap that sometimes exists between researchers and engineers responsible for designing components against fatigue failure. The large portion of papers authored by engineers working for industrial companies illustrates that this objective is being realised. The 21 selected papers provide an up-to-date survey of engineering practice and a preview of design methods that are advancing toward application.

Reliability was selected as a key theme for FD'98. During the design of components and structures, it is not sufficient to combine mean material properties, average usage parameters and pre-selected safety factors. The engineer must also consider potential scatter in material properties, different end users, manufacturing tolerances and uncertainties in fatigue damage models. Judgement must also be made about the consequences of potential failure and the required degree of reliability for the structure or component during its service life.

Approaches to ensuring reliability may vary greatly depending on the structure being designed. Papers in this volume intentionally provide a multidisciplinary perspective on the issue. Authors represent the ground vehicle, heavy equipment, power generation, ship building and other industries. Identical solutions can not be used in all cases because design methods must always provide a balance between accuracy and simplicity. The point of balance will shift depending on the type of input data available and the component being considered.

A large number of people contributed to the success of both the symposium and this publication. The editors gratefully acknowledge the roles played by the scientific and advisory committees, the manuscript reviewers as well as staff members at ESIS, Elsevier and VTT. The greatest thanks, however, is reserved for the authors who have invested countless hours developing the ideas presented here and their care in preparing the papers.

G. Marquis and J. Solin, Editors
VTT Manufacturing Technology

FATIGUE DESIGN AND RELIABILITY IN THE AUTOMOTIVE INDUSTRY

JJ THOMAS, G. PERROUD, A. BIGNONNET, D. MONNET
PSA Peugeot Citroën, Centre SAMM - Chemin de la Malmaison, 91570 Bièvres -France

ABSTRACT

Fatigue assessment of automotive components is performed at PSA using a statistical approach to predict reliability. Calculations provide a lot of information but experimental work is always necessary to determine this reliability. Results exploitation is carried out using the "Stress-Strength interference analysis" method (SSIA). The "Stresses" represent the distribution of the car owners severity, the "Strength" represents the distribution of the fatigue strength of all the components. If the determination of the Strength is easy (fatigue tests results), the application of the "Stress-Strength" Method also requires the knowledge of the owners severity. This information is accessible as statistical distribution through owner enquiries on car usage. On the one hand a "car usage" enquiry allows conditions in which cars are driven (loading, route) to be known. On the other hand an "owner behaviour" enquiry allows the determination for a given use of the distribution of the owners' severity. The fatigue assessment of an engine subframe is used to illustrate the approach.

KEYWORDS

Reliability, Stress-Strength interference analysis, usage enquiries, fatigue, welded components

INTRODUCTION

Fatigue assessment of automotive components, and particularly welded ones, is performed at PSA using a statistical approach to evaluate the reliability. Calculations provide a lot of information but experimental work is always necessary to determine this reliability. Results exploitation is carried out using the « Stress-Strength Interference Analysis » method (SSIA) [1] [2]. The "Stresses" represent the distribution of the car owners severity, the "Strength" represents the distribution of the fatigue strength of all the components. If the determination of the Strength is easy (fatigue tests results), the application of the SSIA method is somewhat difficult due to the lack of knowledge of the owners severity. This information is accessible through owner enquiries on car usage. On the one hand a "car usage" enquiry allows the conditions in which cars are driven (loading, route) to be known, on the other hand an "owner behaviour" enquiry allows you to determine for a given use, the distribution of the owners' severity.

Based upon these principles, the assessment of the individual components and of the good functioning of the whole, rests on two pillars : modelling and testing. This ensemble is enriched by accumulated experience. The keystone of this edifice are the specifications which set the geometrical constraints linked to vehicle architecture and the types of loading encountered in service.

Within this framework, the results of modelling and testing will be accepted in light of objective criteria adapted to each specification, i.e. : fatigue damage, stiffness...

With respect to the problem of the fatigue strength, the first step is to evaluate the loading history the vehicle will undergo during its whole life. In the second step, the loading must be processed in a form acceptable for its use in design offices to perform fatigue assessment calculations.

At this stage, a geometric optimisation can be performed assuming that other specifications are met : stiffness, shock resistance... Finally, testing on components and vehicles, allows the validation of the whole unit to detect possible problems, and to ensure that the level of reliability is satisfactory.

In the following, examples of industrial practice are presented.

LOADING SPECIFICATIONS AND ACCEPTANCE CRITERIA

To clearly understand the strategy for the approach to fatigue strength in the automotive industry, it is necessary to specify what these external loadings are. Two types of loading must be distinguished :

- Loading from the normal use of the vehicle, which does reflect and contain the variability of the customers. The resistance criteria should take into account the scatter due to fabrication.
- Loading from accidental or exceptional situations, which may occur a dozen times (at most) in the life time of the vehicle (for example, release the clutch at full regime, obstacle crossing with blocked wheels...) but in any case, which should not affect passengers security.

As an illustration, let's take the suspension system. The suspension is typically submitted to a great number of loading cycles, which are multiaxial and of variable amplitude.

Maximum loading and fatigue loading

The loading specifications and resistance criteria are based on the stress - strength reliability approach. From measurements performed on customer cars, loading histories at the wheel base are given in the three directions : longitudinal, lateral and vertical (X, Y, Z). A statistical analysis on a large population allows the definition of an "objective customer" with a known severity (mean plus a given number of standard deviations in the case of a gaussian distribution).

The loading histograms corresponding to this objective customer yield the following :

- the maximum values of the loading in each direction, events that could occur a few thousand of times in the life of the vehicle.
- all the load cycles on the whole life of the vehicle (including the maximum values), which will produce the fatigue damage.

On the one hand, these loadings must be withstood without permanent deformation (the material remains in its elasticity range) and on the other hand, they must be withstood without apparent fatigue degradation, (no crack can be detected).

The acceptance of individual components, in validation testing or in predictive calculation is based on the stress - strength reliability approach, as shown in a further paragraph.

EQUIVALENT FATIGUE LOADING

To analyse the loading and to provide useful information to the designers, the fatigue loading recorded on vehicles is transformed into an equivalent loading (couple : forces of constant amplitude - number of cycles), which produces the same fatigue damage that the vehicle will support in its whole life. This fatigue loading corresponds to the objective customer defined in the previous section.

The equivalent fatigue loading cycle which has to be determined is defined by its mean value FATmean and its amplitude FATamp. The usual procedure can be schematically described as follows :

- rainflow counting of the load signal, in each direction (X, Y, Z).

  Two counting methods are widely used for the signal analysis : the level crossing method, or the rainflow counting method. This last method, which is more widely used now, defines cycles which represent hysteresis stress-strain or load-displacement loops. These cycles can be represented in histograms of alternate loading F_{Aj} associated with mean load values F_{Mj} versus number of cycles n_j.

  Cycle counting methods such as rainflow counting are uniaxial. It means that the phase of the X,Y and Z loading directions is lost. Therefore, the most appropriate directions on wich the loading cycle counting are realised have to be identified. This identification is not automatic, it is only based upon the observation of the physical situations. For example, in suspension components, the X direction is mainly loaded when the other directions are nearly constant. It is therefore treated separately, while the two other directions (Y,Z) can be treated together.

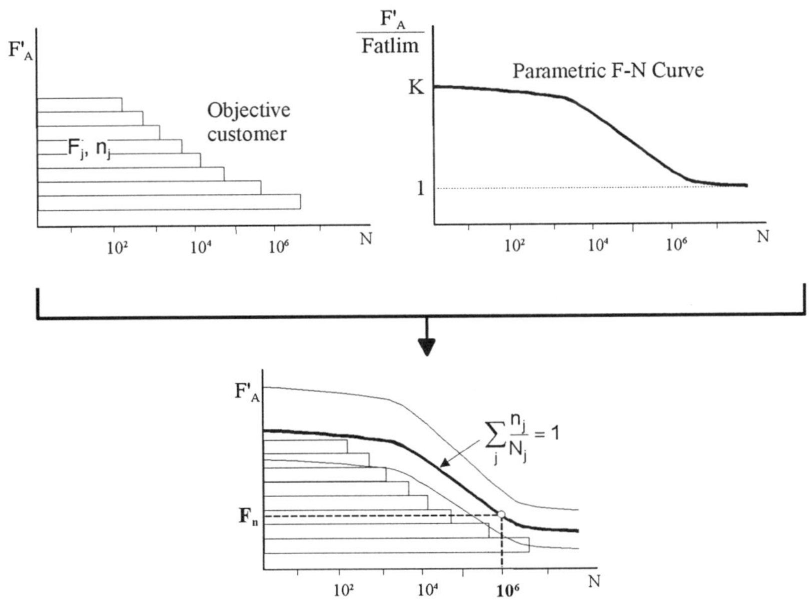

figure 1 : Equivalent fatigue loading

- The global mean value, FATmean, is determined from the whole loading sequence. It is the mean value of all the F_{Mj} value weighted by the associated n_j.

- transformation of each class of cycles with a non zero mean load value ($F_M \pm F_A$), to equivalent purely alternate loading cycles (F'_A), using a parametric GERBER parabola normalized to the fatigue limit seeked for the component : FATLIMi :

$$F'_A = \frac{F_A}{1 - \left(\frac{F_M}{K.FATLIM_i}\right)^2}$$

with K = ratio between the fatigue limit and the ultimate tensile strength of the material considered (typically 2.5 for steels).

- MINER summation is performed with this "objective" histogram and a parametric WHOLER curve normalized to the fatigue limit (i.e. fatigue limit = 1).

  The desired information is the WOHLER curve defined by the value of FATLIMi, which will give a MINER summation of 1 as shown on figure 1 (for details see Morel et al. 1993).

- Any point of this WOHLER curve (couple : force amplitude - number of cycles) can be taken as an equivalent of the fatigue loading experienced by the objective customer. For suspension systems, the equivalent is usually defined at 10^6 cycles.

IN SERVICE LOADING

This approach aims at the knowledge of the statistical distribution of the equivalent fatigue loading. The histograms used in this approach include the high level loading cycles (for which partial damage may be relatively high), but does not include accidental loadings, wich are supposed to be very infrequent (less than 10 in the whole life of the vehicle).

The knowledge of these values allows the statistical distribution of the damaging loads to be determined. They are associated to :

- the car usage : all the owners do not use their car in the same conditions ; the car is more or less loaded, roads are different (highway, city, uneven road, mountain, ...)
- the owner behaviour : the "driving style" can be sporty or quiet, ...

The whole approach is based upon the calculation of an equivalent fatigue alternate loading of constant amplitude defined for 10^6 cycles which represent a number of kilometres covered by the car in its whole lifetime [3]. This calculation is made using a Rain-Flow count, Whöler or Basquin curves, the Gerber parabola and the Miners' rule.

Car usage

General enquiries provide mean parameters for a given population but information on scatter is scarce. For example, in France the mean occupation of a vehicle is 1.8 people and the mean percentage of highway driving is 26%.

More precise enquiries are necessary.

It consists of enquiries performed with one thousand car owners comparable to the one studied. The present example is for the M1 segment. For each owner questioned, the spread of the car usage is obtained in terms of the typical road and load carried. The range of utilisation obtained is assumed to be representative of the entire owners.

An owner is therefore characterized in terms of car usage by the relative percentage of kilometres driven with various load states, and for each load state, the relative percentage of kilometres covered on each road type. For example if 3 load states and 4 road types are taken into account, the car usage of an owner takes the shape given in table 1.

The entire owners, as described above constitute the target population. Each owner j interviewed during the car enquiry is therefore characterized by a vector U_j ; the sum of these representative owners gives a matrix [U].

Owner behaviour

The owner behaviour is defined by the way he drives, i.e. the way the structures ares loaded in the various situation of road and load states. Numerically, this behaviour is represented by « elementary 1 kilometer Rain-Flow matrices » $[h_{jkl}]$ which are recorded in each situation : j, k, l represent the identification of the driver, the load state, the road type respectively.

A car equiped with sensors to measure the desired loading information is placed at representative owners disposal. They must drive on a predetermined run. The run containing the different road types is performed at several load states. The number of load states and the number of routes are not limited. Nevertheless one should ensure coherence between the two enquiry types. In our example the strain evolution is recorded for each owner on each road type. These records correspond to a owner which uses a determined route and for a given load state. With the sum of each measurement obtained, files contain a matrix $[h_{jkl}]$

table 1 : Car usage description for two owners

		Owner (j)	1	2
% kilometers without load			27	15
	elementary %	Motorway	10	25
		Good road	25	12
		Mountain	40	50
		City	25	13
% kilometers with half load			58	35
	elementary %	Motorway	5	16
		Good road	30	24
		Mountain	30	40
		City	35	20
% kilometers full load			15	60
	elementary %	Motorway	15	18
		Good road	25	42
		Mountain	40	10
		City	20	30

table 2 : Owner behaviour with different car usage

	Usage $\mathbf{U_1}$ Road % for total kilometers	Usage $\mathbf{U_2}$ Road % for total kilometers
Load state 1 (without load)	27	15
% Motorway	$[h_{11A}]$	$[h_{21A}]$
% Good road	$[h_{11R}]$	$[h_{21R}]$
% Mountain	$[h_{11M}]$	$[h_{21M}]$
% City	$[h_{11V}]$	$[h_{21V}]$
Load state 2 (half load)	58	35
% Motorway	$[h_{12A}]$	$[h_{22A}]$
% Good road	$[h_{12R}]$	$[h_{22R}]$
% Mountain	$[h_{12M}]$	$[h_{22M}]$
% City	$[h_{12V}]$	$[h_{22V}]$
Load state 3 (full load)	15	60
% Motorway	$[h_{13A}]$	$[h_{23A}]$
% Good road	$[h_{13R}]$	$[h_{23R}]$
% Mountain	$[h_{13M}]$	$[h_{23M}]$
% City	$[h_{13V}]$	$[h_{23V}]$

Enquiries exploitation

The enquiries exploitation aims at the determination of the whole customer Rain-Flow matrix distribution. This is achieved through the assumption that usage and owner behaviour are two independent parameters.

Therefore, it consists of obtaining the Rain-Flow matrix [H] for each recorded information corresponding to a driver j who should for example drive his car following the spread given by $[U_k]$, during the whole lifetime of the car.

This combination is made proportionally to the length of the various roads corresponding to the $[h_{jkl}]$

matrix and to those indicated in the spread vector U_j. An example is given in table 2

The Rain-Flow matrix for the drivers are therefore :

$$\text{driver 1 : } [H_{11}] = N\left(\left([h_{11A}] \text{ x } 0.27 \text{ x } 0.1\right) + \left([h_{11R}] \text{ x } 0{,}27 \text{ x } 0{,}25\right) + \ldots\right)$$

$$\text{driver 2 : } [H_{22}] = N\left(\left([h_{21A}] \text{ x } 0.15 \text{ x } 0.25\right) + \left([h_{21R}] \text{ x } 0{,}15 \text{ x } 0{,}12\right) + \ldots\right)$$

virtual driver using his car like driver 1, and behaving like driver 2 :

$$[H_{12}] = N\left(\left([h_{21A}] \text{ x } 0.27 \text{ x } 0.1\right) + \left([h_{21R}] \text{ x } 0{,}27 \text{ x } 0{,}25\right) + \ldots\right)$$

N is the number of kilometres for the whole lifetime of the car.

To define the distribution of equivalent fatigue loading, simulations of about 10 owners are performed with the data coming from the enquiries The equivalent fatigue damage condition is determined for each matrix [H].

This calculation is performed for all the possible combinations of owners behaviour and car usage. The entire results which correspond to the simulation of 10 000 to 50 000 virtual owners, allows the distribution of the equivalent fatigue condition of actual car utilisation to be determined. This is called the distribution of the owners' severity or the stress distribution for the measured value, figure 2.

These distributions can reasonably be described by a Normal law. The stress distribution is therefore defined by its mean value μ_c and its standard deviation σ_c.

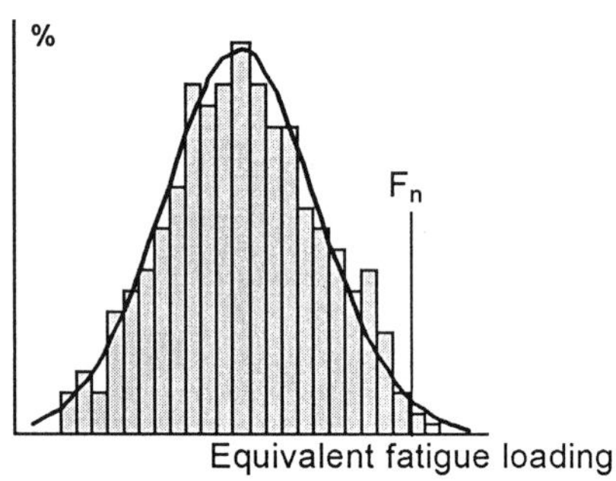

figure 2 : Equivalent fatigue condition distribution

Determination of the objective owner

From the stress distribution, an objective owner F_n is defined such that :

$$F_n = \mu_c + \alpha\, \sigma_c \qquad (1)$$

The probability of finding a more severe owner is given by the normal law. For example if $\alpha = 4.1$, therefore :

$$\text{Prob}\left(\text{severity} > F_n\right) = 1/50000 \qquad (2)$$

The fatigue calculations and testing are performed with this level of severity.

THE "STRESS-STRENGTH INTERFERENCE ANALYSIS" FOR A RELIABLE DESIGN

Once the distribution of stress is know, two more inputs are necessary to achieve the reliable design. First we define the risk R of failure in service, (i.e. the predictive reliability F = 1-R). This risk represents the probability that a customer would « met » a component too weak for him. The reliable design is obtained by placing the strength distribution on the right side of the stress distribution in order to respect the risk R. It shows that the knowledge of the strength relative scatter is also necessary. Since the component is at the design stage, its relative scatter is unknown.

The stress, random variable C represents the distribution of the stress of all the owners. The strength, random variable r, represents the distribution of the strength of the entire components. figure 3 illustrates those two distributions and the position of the testing reference.

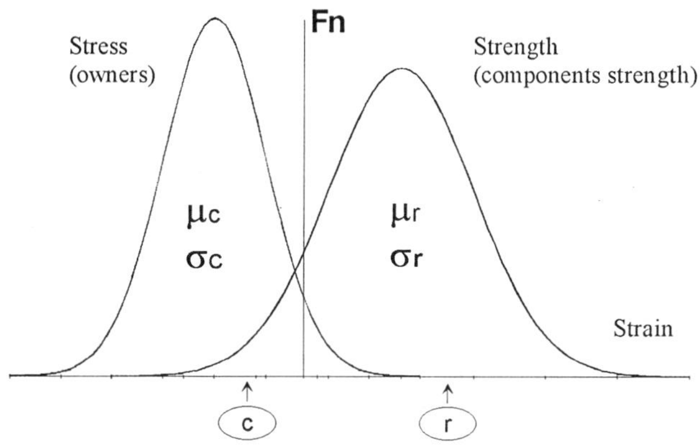

figure 3 : Stress Strength Interference Analysis method illustration

As "r" and "c" are Normal laws the random variable Z = r - c, also follows a Normal law characterized by :

$$\mu_Z = \mu_r - \mu_c \quad \text{and} \quad \sigma_Z = \sqrt{\sigma_r^2 + \sigma_c^2}$$

Considering the centered reduced variable $u = \dfrac{z - \mu_Z}{\sigma_Z}$, the risk R is such that

$$R = \text{Prob}\,(z < 0) = \text{Prob}\left(u < -\frac{\mu_Z}{\sigma_Z}\right)$$

this can be derived from the Normal law tables. Actually the "Strength" parameters μ_r and σ_r are estimated with a confidence level γ by two values m_r and s_r obtained from a limited number of components. Therefore, the risk R depends on the chosen confidence level γ and of the number N of tested components.

Usually, the number of tested components ($N \geq 8$) allows a reasonable estimation of the mean value and $\mu_r = m_r$ but on the other hand the standard deviation must be corrected by χ^2 :

$$\sigma_r = s_r \sqrt{\frac{\nu}{\chi^2_{\gamma(\nu)}}}$$

with $\nu = N - 1$ and $\chi^2_{\gamma(\nu)}$ the value for the chosen iso-probability at N-1 degrees of freedom

If $N = \infty$ then $\sigma_r = s_r$

NB : m_r and s_r are homogeneous to the applied Forces and they can be normalized by the testing reference : $m_r^* = \frac{m_r}{F_n}$ and $\sigma_r^* = \frac{S_r}{F_n}$

For an easier analysis of the results a relative scatter parameter is introduced in each of the distributions : $\rho = \frac{\sigma_c}{\mu_c}$ and $q = \frac{S_r}{m_r}$, ρ is representative of the shape of the stress distribution, while q is characteristic of a component family and its fabrication process. Therefore the risk can be written :

$$R = \text{Prob}\left(u < -\frac{\mu_Z}{\sigma_Z}\right) = f\left(\alpha, \rho, m_r^*, q, N, \gamma\right)$$

The risk is calculated with the following procedure :

- values are normalised by F_n
- the data are : α, ρ, m_r^*, q, N, γ
- Calculation : $$\frac{\mu_Z}{\sigma_Z} = \frac{m_r^* - \dfrac{1}{1+\alpha\rho}}{\sqrt{\left(qm_r^*\right)^2 \dfrac{N-1}{\chi^2_{\gamma(N-1)}} + \left(\dfrac{\rho}{1+\alpha\rho}\right)^2}} \qquad (3)$$
- read $R = \text{Prob}\left(u < -\frac{\mu_Z}{\sigma_Z}\right)$ on a Normal law table.

The estimation of the component strength scatter (σ_r) is extremely important. An error of 20% on S_r could bring a factor of 10 on the estimation of the Risk R. The lower the number of tested components, the higher the value of σ_r and the greater the risk value. Therefore it is interesting to work with the parameter q which characterises the component and its fabrication process.

A data base derived from a large number of tests performed on components or specimens can provide a reliable value of the relative scatter parameter q. In this case, only the determination of the mean value of the "Strength" distribution is necessary. This can be carried out with a limited number of components, ten for example. It is no longer necessary to take into account the number of tested components (m_r and q are considered to be representative of the whole components). The calculation of the risk R is more precise and is not penalized by a correction due to the number of tested components.

APPLICATION TO AN ENGINE SUBFRAME

The fatigue strength of welded components can be rather scattered depending on the welding process control. Within the framework of an automotive project, in the development phase, this scatter is not easy to access because prototype components are fabricated manually in small quantities. The true scatter becomes measurable when mass production is reached (thousands of components / day). For welding to be economically competitive with regard to the other possible processes, forge or foundry, it is necessary to provide the fabricants with the analysis tools in order to guarantee a low scatter of the components fatigue strength.

Let's take an example as an illustration :

- a specification imposes a failure risk $R \leq 10^{-6}$ for a component.

- The designer has the choice between two fabrication processes. One process ensures a low scatter fatigue strength characterized by q = 0,06 ; the other process has the advantage of being cheaper and allows a lightweight design for the same strength but the fabrication scatter is larger, q = 0,10.

With the first solution the design objective is reached with a mean strength value $m_r^* = 1.25$; with the second solution, to guarantee the risk level the mean strength value must be $m_r^* = 1.55$.

Economical and component weight considerations evaluated at a first glance are in this case inversed after the statistical analysis, revealing that the choice of the second process finally leads to a 25% increase in the component mass.

That example is often encountered when comparing forging or welding process on mechanical components.

The objective mean strength value which should be reached, depending on *q* and *N*, is given on figure 4. This figure shows how important the relative scatter of the process is.

Let's take an example as an illustration :

Let us compare two cases for a specification imposing a failure risk $R \leq 10^{-6}$ for a component. The curves on figure 4 show that :

- If the relative scatter *q=0,08* and the number of tested pieces is large, the design objective is reached with a mean strength value $m_r^* = 1{,}27$.
- If the relative scatter *q=0,09* and the number of tested pieces is equal to 10, the design objective is reached with a mean strength value $m_r^* = 1{,}55$.

We can apply this method to a welded engine subframe (see figure 5). Data collected by Fayard [4] on elementary structures lead to a relative scatter *q=0,08*. This value should be associated with the welding process, and therefore can be used for our application.

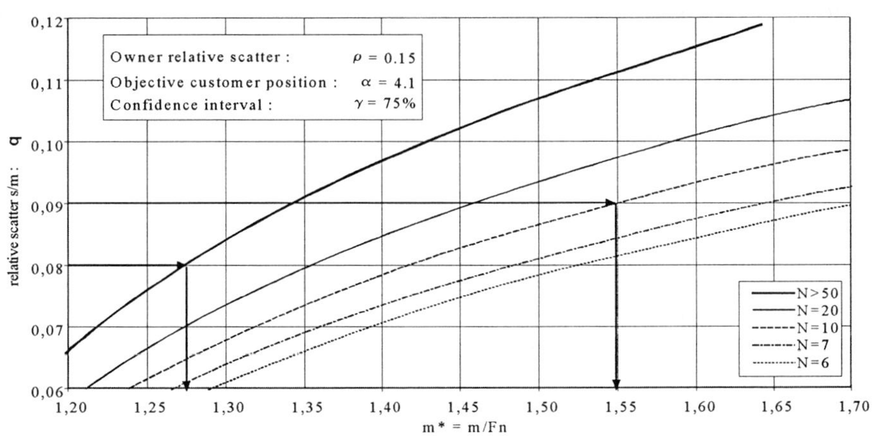

figure 4 : determination of the objective mean stength from the relative scatter and the number of tested pieces.

Tests have been realised at different load levels on the suspension understuctures. The loads and their associated lifetimes have been converted by Miner's law to loads leading to 10^6 cycles.

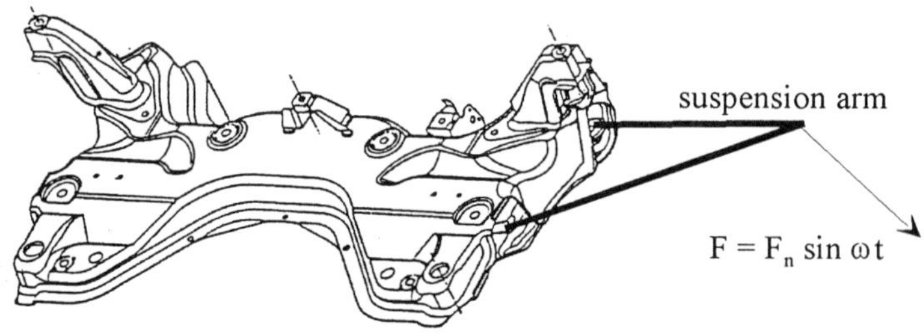

figure 5 : engine subframe used for experimental and numerical works

These load levels were introduced in numerical simulations (see figure 6) using the meshing rules defined by Fayard [4]. For a given load level, the maximum principal stress depends on the observed area. For each defect, the experimental points given by the lifetime and the maximum principal stress are converted into equivalent values of maximum principal stresses leading to 10^6 cycles using the fatigue curves of Fayard. Plotting the results for each maximum principal stress at this number of cycles illustrates the observed scatter of experimental results. The relative scatter $q=0,08$ associated with the welding process allows a gaussian curve to be constructed at 10^6 cycles. Figure 7 shows that the experimental dispersion observed on the engine subframe is in good agreement with the dispersion assessed from elementary structures.

In the classical design procedure, an acceptance criteria is necessary to represent the objective m* and therefore guarantee the risk R. For $q=0,08$ figure 4 shows that m* should be equal to 1,27. The correct isoprobability curve (figure 7) is situated at (1,27-1)/0,08 ~ 3,4 std deviations. This number of standard deviations is used to determine the acceptance criteria of the finite element fatigue analysis. In the presented exemple, for a equivalent fatigue loading at 10^6 cycles, and for a risk $R=10^{-6}$, the loading level is F_n (defined by equation 1), and the corresponding acceptance criteria is obtained on figure 7 from the curve at the mean value minus 3,4xstd deviations.

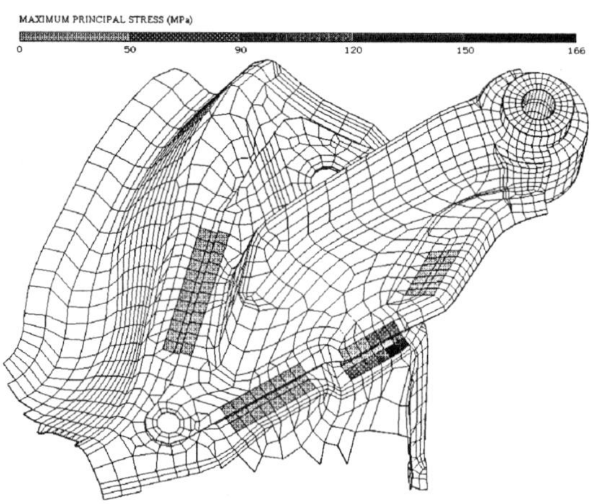

figure 6 : exemple of numerical results for the welded joints of the engine subframe

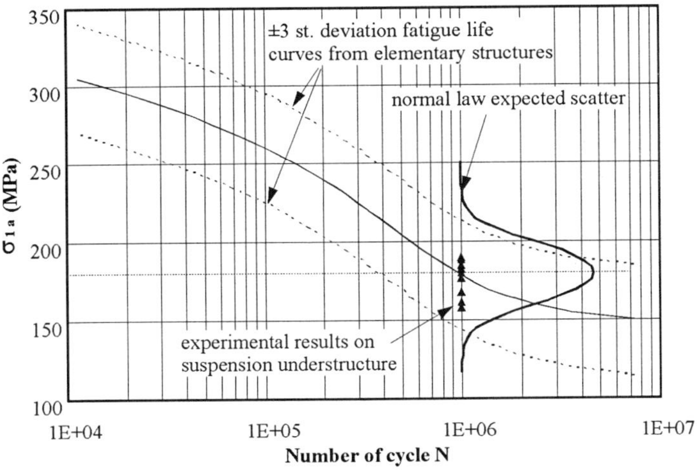

figure 7 : comparison between experimental and expected scatter on the engine subframe

CONCLUSIONS

This work presents the analysis of the fatigue strength of automotive components in service using the "Stress-Strength" analysis.

The major difficulty is the definition of the service loading which demands long and costly statistical analysis of car usage and owner behaviour. The determination of the components fatigue strength distribution is easier but its accuracy depends strongly on the number of components tested. In practice, it is often difficult to test many components (cost at prototype stage, lengthy testing). Therefore data bases are built to appreciate the relative scatter parameters typical of each fabrication process. This allows the statistical evaluation of the risk to be improved while reducing the number of tests.

It is important to note that the "Stress-Strength" method points out the paramount importance of the relative scatter parameter of fatigue strength. From experience it is known that the value of this parameter can be high on welded components.

REFERENCES

1. F. Morel, J. Mercier, G. Catherin, A. Bignonnet, J. Petit, Analyse du comportement en fatigue de composants par l'approche du chargement équivalent,*in Sollicitation en service et comportement en fatigue*, Paris SF2M, 1993.

2. D. Kececioglu, Reliability analysis of mechanical components and systems, *Nuclear Engineering and Design,* **19**, 259-290 (1972).

3. C. Marcovici, J. C. Ligeron, . (PIC Edition, Genève, 1974) pp. 87-107.

4. J. L. Fayard, A. Bignonnet, K. Dang-Van, Fatigue design of welded thin sheet structures, G. Marquis, J. Solin, Eds.,*in Fatigue design 95*, Helsinki 1995.

RELIABILITY BASED FATIGUE DESIGN OF MAINTAINED WELDED JOINTS IN THE SIDE SHELL OF TANKERS

C. Guedes Soares and Y. Garbatov

Unit of Marine Technology and Engineering
Technical University of Lisbon, Instituto Superior Técnico,
Av. Rovisco Pais, 1096 Lisboa, Portugal

ABSTRACT

The present work deals with the application of reliability based techniques to the design welded joints subjected to the process of crack growth and repair. A formulation is presented for the assessment of the fatigue damage and of the reliability of the side shell of a ship hull structure. The potential cracks are considered to occur in the side shell, in the connections between longitudinal stiffeners and transverse web frames. The model accounts for the crack growth process applying linear elastic fracture mechanics. The long-term stress range acting on the elements is defined as a function of the local transverse pressure of internal cargo and outside water, combined with the stresses resulting from the longitudinal bending of the hull namely the a combination of horizontal and vertical bending moments. The fatigue reliability is predicted by a time variant formulation and the effects of maintenance actions in the reliability assessment are shown.

KEYWORDS

Fatigue, crack growth, inspection, reliability, and maintenance.

INTRODUCTION

The developments in the understanding of the nature of the loading, the intensified use of new materials with higher strength capacity, the frequent utilisation of refined analyses in the design processes have made possible an optimisation of structures in general and in particular during the last decade. The production of more economical structures has however made them more prone to the effect of the strength degradation phenomena such as fatigue and corrosion.

Fatigue design is one of the most complicated problems in engineering, especially for the structural components subjected to stochastic loading and predicting a component reliability under the fatigue failure mode is generally difficult, not only, because of the difficulty in describing the mechanics of fatigue failure, but also because of the complexity of the reliability model.

Ship structures may contain randomly distributed fabrication imperfections due to material and workmanship quality, and have to stand various types of loads which themselves are characterised probabilistically. Therefore, the rational design of such structures should be based on first principles and probabilistic descriptions of loads and strength.

Ship structures should be designed considering the demands and changes that occur over their operational life. This requires addressing maintenance considerations at the design stage. This means that the techniques should support the reassessment of the structural condition at any time, which can be done with the model presented here in a manner similar as other computational are used for design.

The fatigue reliability of a joint in a ship structure has been studied in detail in [1] while the time variation of ship reliability due to fatigue has been presented in [2]. The study of the fatigue reliability of longitudinal members in the ship structure under longitudinal bending has been conducted in [3] considering the overall effect of a random number of cracks occurring during the life of the ship and maintenance action. It was considered that the growth of a crack decreases the net area of stiffeners or plating that contribute to the longitudinal strength. The overall effect of the simultaneous action of a random number of cracks is modelled as a decrease in the net area of the midship section, which resist the longitudinal bending of the hull.

The effect of the vertical distribution of pressures in the side shell has been considered recently in [4-6] where the S-N approach has been used and only fatigue reliability of unrepaired structures are discussed. Linear elastic-fracture mechanics was adopted for fatigue reliability of ship structures in [7] where an approach is presented that incorporates the effect of inspections with an uncertain outcome, and of repair of the detected cracks. The problem treated considers the fatigue reliability of joints in the side shell of ships, taking into account the combined effect of the pressure loading in the side shell and the longitudinal bending of the hulls.

The present paper addresses the problem of joint design based on a reliability based formulation. It is shown that applying the present approach, the relative number of replaced elements in the side shell varies as a function of vertical position and has a maximum chose to the waterline

It is demonstrated then how one can redesign the structure by redistributing the material in the longitudinal stiffeners so as to reduce the total number of replaced elements, with the constraint that the overall section modulus of the midship section can not be changed. This approach corresponds to the normal design practice in which the midship section is designed based on longitudinal strength considerations and in further iterations other aspects such as fatigue are considered.

It is also demonstrated how this formulation can be used for reliability based maintenance planning and in particular how to vary the inspection interval in order to vary the maximum number of repaired elements, for the same level of reliability. Alternatively, for fixed inspection intervals it is shown how the initial crack size, detectable crack size, time interval between inspection, average period of the sea state of the minimum reliability and the number of repaired elements vary along the ship life. It is demonstrated that keeping certain value of minimum reliability reflects to minimum number of replaced elements.

LOADING OF THE SHIP STRUCTURE

In the evaluation of the dynamic stress levels at a local structure, both the global ($\Delta\sigma_M$) and local ($\Delta\sigma_p$) dynamic stress components need to be considered. The global stress components include wave-induced vertical and horizontal hull girder bending stresses. The local stress component result from the external sea pressure and the pressure loads from internal cargo. For each loading condition, the local stress components need to be combined with the global stresses.

The long-term distribution of global stress amplitudes may be estimated using the long term frequency of occurrence of different sea states, where each sea state is described by a significant wave height and zero crossing period. The long-term distribution of the response is established as the weighted sum of the individual short term response distributions over all the sea states and heading directions, weighted with the relative occurrence rate of response cycles, [8].

The combined total response in terms of the stress range $\Delta\sigma$ resulting from the combination of global $\Delta\sigma_M$ and the local $\Delta\sigma_p$ stress ranges may be given as the largest of $[\Delta\sigma_M + 0.6\Delta\sigma_p]$ and $[0.6\Delta\sigma_M + \Delta\sigma_p]$:

$$\Delta\sigma = max\{[\Delta\sigma_M + 0.6\Delta\sigma_p] \cup [0.6\Delta\sigma_M + \Delta\sigma_p]\} \tag{1}$$

according to the rules of Classification Societies, e.g. [9]. This code specification of the design load corresponds to the application of the Turkstra rule, which has been adopted in many occasions to prescribe design rules, [10].

The stress range response is estimated from the vertical ($\Delta\sigma_{WV}$) and horizontal ($\Delta\sigma_{WH}$) wave induced hull girder bending stresses:

$$\Delta\sigma_M = \sqrt{\Delta\sigma_{WV}^2 + \Delta\sigma_{WH}^2 + 2\rho_{VH}\Delta\sigma_{WV}\Delta\sigma_{WH}} \tag{2}$$

where ρ_{VH} is the correlation coefficient.

The correlation coefficient between vertical and horizontal stresses (ρ_{VH}) is given:

$$\rho_{VH} = \cos(\varepsilon_H - \varepsilon_V)\cos(\omega t - \varepsilon) \tag{3}$$

where phase angle (ε) is determined from:

$$\varepsilon = arctg\left[\frac{\Delta M_{WV}\sin(\varepsilon_V) + \frac{W_y}{W_z}\Delta M_{WH}\sin(\varepsilon_H)}{\Delta M_{WV}\cos(\varepsilon_V) + \frac{W_y}{W_z}\Delta M_{WH}\cos(\varepsilon_H)}\right] \tag{4}$$

It can be seen that the combined stress amplitudes, which result from the vertical and horizontal induced bending moments are dependent on the ratio W_y/W_z and on the phase difference $(\varepsilon_H - \varepsilon_V)$. There are many places in the literature where it is noted that $(\varepsilon_H - \varepsilon_V)$ is mainly a function of the ship heading and is dependent on the ratio between the wavelength and the length of the ship (λ/L), [7].

The combined local stress range is estimated assuming an average long term distribution between external sea pressure induced stress amplitude (σ_e) and inertial pressure induced stress amplitude (σ_i):

$$\Delta\sigma_p = 2\sqrt{\sigma_e^2 + \sigma_i^2 + 2\rho_{ei}\sigma_e\sigma_i} \tag{5}$$

An adequate approximation for the long-term distribution of wave induced loading can be described by the Weibull distribution, [8]:

$$F(\Delta\sigma) = 1 - \exp\left[-\left(\frac{\Delta\sigma}{q} \right)^h \right] \tag{6}$$

The shape parameter h depends on the parameters of the ship, the location of the detail, and the sailing routes during the design life.

RELIABILITY OF A CRACKED ELEMENT WITHOUT MAINTENANCE

To predict the fatigue life crack propagation the Paris-Erdogan equation has been adopted:

$$\frac{da}{dN} = C\Delta K^m, \Delta K > K_{th} \tag{7}$$

where a is the crack size, N is the number of cycles, ΔK is the stress range intensity factor, C and m are material parameters and ΔK_{th} is the stress range threshold intensity factor. The stress intensity factor is given by the follow equation:

$$\Delta K = \Delta\sigma \, Y(a)\sqrt{\pi a} \tag{8}$$

where $\Delta\sigma$ is the stress range and $Y(a)$ can be expressed as [14]:

$$Y(a) = F_e F_s F_w F_g \tag{9}$$

where F_e, F_s, F_w and F_g are a crack shape, a free surface, a finite width and a stress gradient correction factor.

If $Y(a) = Y$ is a constant, $N = \nu_o t$ and after substitution of (9) into (8) and integration of Eqn (7) one obtains:

$$a(t) = \left[a_o^{\,1-\frac{m}{2}} + \left(1 - \frac{m}{2}\right) C \, \Delta\sigma^m \, Y^m \, \pi^{\frac{m}{2}} \, \nu_o t \right]^{\frac{1}{1-\frac{m}{2}}}, \quad m \neq 2 \tag{10}$$

where ν_o is the mean upcrossing rate and t is the time, $\Delta\sigma^m$ is the mth moment of the stress range.

The time to crack initiation is modelled by a Weibull distribution, which was recommend in [11].

The limit state for a cracked element of low carbon mild steel may be defined as:

$$a_{cr} - a(t) \leq 0 \tag{11}$$

where a_{cr} is the critical crack size, and $a(t)$ is the crack size, which depends of time. The critical crack size is defined here as a percentage of the height of the stiffener or the breadth on plate element.

Failure will take a place if the stress time history upcrosses the limit $\zeta(t)$, which can be written as:

$$\zeta(t) = \left(\sigma_1 - b_1 \nu_o \sigma_1 t\right)^{1/(m-2)} \tag{12}$$

where the coefficients included in Eqn (12) are:

$$\sigma_1 = \frac{K_{cr}}{Y\sqrt{\pi a_o}}, \quad b_1 = \frac{m-2}{2} Y^m \pi^{m/2} a_o^{m/2-1} C \Delta\sigma^m \tag{13}$$

The probability that the stress would exceed $\varsigma(t)$ during the period of the time $[0,T]$, i.e., the probability of failure, is [12]:

$$P_f(T) = 1 - \exp\left[-\int_o^T \nu\left[\varsigma(t)\right] dt\right] \tag{14}$$

where $\nu\left[\varsigma(t)\right]$ is the mean upcrossing rate of the threshold $\varsigma(t)$.

The probability of failure after crack initiation is written as:

$$P_a(T) = 1 - \exp\left\{-\frac{\gamma_L}{b_1\sigma_1}\left[\exp\left(\frac{\sigma_1 - b_1\sigma_1\nu_o T}{\gamma_L}\right) - \exp\left(-\frac{\sigma_1}{\gamma_L}\right)\right]\right\} \tag{15}$$

Making t=0 defines the zero upcrossing rate before cracks have initiated. The probability of failure before crack initiation is given:

$$P_b(T) = 1 - \exp\left\{-\exp\left[-\frac{\sigma_1}{\gamma_L}\right]\nu_o T\right\} \tag{16}$$

The probability of non failure before a crack initiation, $R_b(t)$ is written by:

$$R_b(t) = 1 - P_b(t) \tag{17}$$

and the probability of non-failure after crack initiation $R_a(t)$ is described by same way.

Since the time to crack initiation is a random variable, the conditional reliability of the element with a crack may be expressed as follows:

$$R(T) = \int_o^T R(t|t_i) f_{t_i}(t) dt \tag{18}$$

where $R(t|t_i)$ is the reliability under the condition that the crack has initiated at time t and $f_{t_i}(t)$ is the probability density function of the time to crack initiation. If $R(T)$ is the reliability in the service life, $[0,T]$ the following equation can be written:

$$R(T) = \left[1 - F_{t_i}(T)\right] R_b(T) + \int_o^T f_{t_i}(t)\, R_b(t)\, R_a(T-t)\, dt \tag{19}$$

The first term is the probability of non-failure under the condition that the crack is not initiated during the service time $[0,T]$. The second term is the probability of non-failure under condition that the crack is initiated during the service time $[0,T]$.

RELIABILITY OF A CRACKED ELEMENT ACCOUNTING FOR INSPECTION AND REPAIR

Inspections are routinely made for structures in service and they may result in the detection or non-detection of the cracks. The size of a detected crack is measured by a non-destructive method. For welded structures, cracks are generally assumed to be present after fabrication. Fatigue damage is described by a fatigue crack size that increases with time.

A purpose of periodic inspections is to detect the fatigue cracks. It is assumed that if a fatigue crack is detected, it is repaired to its original condition (a_o) which increases the reliability of the ship's structure.

The inspection quality depends on the ability of detecting the crack and of quantifying its size. In principle each detection technique will have a limit size of detection ($a_{d,o}$), under which cracks will not be identified. The inspection quality may be described by the probability of detection as suggested in [13]:

$$P_d(a) = \begin{cases} 1 - \exp\left[-\dfrac{a - a_{d,o}}{\lambda_d} \right], & if \quad a > a_{d,o} \\ 0 & , if \quad a \le a_{d,o} \end{cases} \tag{20}$$

The inspection quality is characterised by the parameter (λ_d) which has values between 0 and ∞. The smallest limit corresponds to a perfect inspection and $\lambda_d = \infty$, means that the structure has not been inspected. It should be noted that the real measurements of crack size usually involve a large scatter on the crack detection, which is reflected in the different mathematical models describing $P_d(a)$. This uncertainty affects very much the reliability as shown in [14]. In [15] is also showed that the choice of inspection method could have a noticeable effect on the probability of detection. Detailed analysis of the problem can be found in [16].

The structural elements are separated into two groups, including respectively the elements, which were and were not repaired at the time of last inspection T_j. The reliability $R^r_{l,j+1}(t)$ in the service interval $\left[T_j, T_{j+1}\right]$ of an element with a crack that is repaired at the time of the last inspection, can be written as follows:

$$R^r_{l,j+1}(t) = \left[1 - F_{t_i}\left(T_{j+1} - T_j\right)\right] R_b\left(T_{j+1} - T_j\right) + \int_{T_j}^{T_{j+1}} f_{t_i}\left(t - T_j\right) R_b\left(t - T_j\right) R_a\left(T_{j+1} - t\right) dt \tag{21}$$

Using the third axiom of probability theory, (e.g. [17]) the probability of non-failure in the time interval from the last repair T_r up to T_{j+1} can be obtained as:

$$P\left\{C_{T_r,T_j} \cap C_{T_j,T_{j+1}}\right\} = P\left\{C_{T_r,T_j} \middle| C_{T_j,T_{j+1}}\right\} P\left\{C_{T_j,T_{j+1}}\right\} \tag{22}$$

where $C_{T_r,t}$ is probability of non-failure in the time interval $\left[T_r, t\right]$, where $t \in \left[T_j, T_{j+1}\right]$ and C_{T_r,T_j} is probability of non-failure in the time interval $\left[T_r, T_j\right]$. The case $C_{T_r,t}$ includes probability of non-failure C_{T_r,T_j} because $t \geq T_j$. The probability of non-failure in the time interval $\left[T_j,t\right]$, where $t \in \left[T_j,T_{j+1}\right]$ is written as $C_{T_j,t}$.

The Eqn. (22) may be rewritten as a definition of the conditional probability of non-failure in the service interval $\left[T_j,t\right]$, as follows:

$$P\left\{C_{T_j,t}\right\} = \frac{P\left\{C_{T_r,T_j} \cap C_{T_j,t}\right\}}{P\left\{C_{T_r,T_j} \middle| C_{T_j,t}\right\}} \tag{23}$$

However, since the left hand side of Eqn. (23) is equal to the probability of non-failure in the element that is not repaired at the time of the last inspection T_j then:

$$R^{nr}_{I,j+1}(t) = P\left\{\, C_{T_j,t}\right\} \tag{24}$$

The derivation of Eqn (24) and its full expression are too long to be included here but can be found in [7].

The Eqns (21) and (24) include all possible cases of the cracked element, including the states of the crack initiation, the crack propagation, the crack detection, and the crack repair. At the last state, the element is repaired to its original condition and the crack life starts again.

RELIABILITY OF LOCAL SHIP STRUCTURE

Consider a series system with $i = 1,2...m$ series components. The safety margin of the ith component is denoted as M_i, which is given as:

$$M_i = a(t) - a_{cr} > 0 \tag{25}$$

The system probability of failure can be defined as:

$$P_{sys} = P\left(\bigcup_{i=1}^{m} M_i \leq 0\right) \tag{26}$$

The additional information can be included for the events of no crack detection ND_j, the events of crack detection D_j and the events of repair, L_j which are presented as:

$$ND_j = a(t)_j - a_{d,j} > 0 \text{ where } j \in [1, m_{nd}] \tag{27}$$

$$D_j = a(t)_j - a_{d,j} \leq 0 \text{ where } j \in [1, m_d] \tag{28}$$

$$L_j = a(t)_j - a_{l,j} \leq 0 \text{ where } j \in [1, m_l] \tag{29}$$

where $a(t)_j$, $a_{d,j}$ and $a_{l,j}$ are respectively the crack size at certain point of time in the component *j*, the detectable crack size and the crack size which is repaired. The number of events related with crack non-detection, crack detection and repair are m_{nd}, m_d and m_l respectively.

The system probability of the repaired structure is expressed as:

$$P_{sys|ND,D,L} = P\left(M_{sys} \leq 0 \middle| \bigcap_{j=1}^{m_{nd}} ND_j \leq 0 \bigcap_{j=1}^{m_d} D_j = 0 \bigcap_{j=1}^{m_l} L_j = 0 \right) \tag{30}$$

The methods developed for a component can be applied directly to the system case when the equivalent system safety margin is M_{sys}.

To compute the probability of a union of elements, or a series system, one approach is to express it in terms of an intersection of complementary events that can be related to the reliability index by:

$$P\left(\bigcup_{i=1}^{m} M_i \leq 0\right) = 1 - P\left(\bigcup_{i=1}^{m} M_i \geq 0\right) = 1 - \Phi_m(\beta, \rho) \tag{31}$$

where $\beta = \{\beta_i\}$ is the vector of component reliability indexes, $\rho = \{\rho_{ik}\}$ is the vector of correlation coefficients and *m* is total number of components.

Another approach is based on the exact expansion:

$$P\left(\bigcup_{i=1}^{m} M_i \leq 0\right) = \sum_{i=1}^{m} P(M_i) - \sum_{i}^{m}\sum_{i<j}^{m} P(M_i \cap M_j) + \sum_{i}^{m}\sum_{i<j}^{m}\sum_{i<j<k}^{m} P(M_i \cap M_j \cap M_k) - \ldots (-1)^{m+1} P\left(\bigcap_{i=1}^{m} M_i\right) \tag{32}$$

which leads to the upper and lower bounds of Ditlevsen, [18] to the system probability of failure.

The reliability of the structure can be related to the generalised index of reliability which is calculated from a multinormal distribution Φ, [19]. Under this assumption the reliability index (β) can be related with the probability of failure by:

$$P_{sys|ND,D,L} \approx 1 - \Phi_{sys}(\beta, \rho) \tag{33}$$

The first order representation requires the evaluation of the multinormal integral:

$$\Phi_{sys}(\beta, \rho) = \int_{-\infty}^{\infty} \phi(\vartheta) \prod_{i=1}^{m} \Phi\left(\frac{\beta_i - \sqrt{\rho}\vartheta}{\sqrt{1-\rho}} \right) d\vartheta \tag{34}$$

When the probability of failure of components is not correlated, the system probability of failure can be written as:

$$P_{sys|ND,D,L} \approx 1 - \prod_{i=1}^{m} \Phi(\beta_i) \tag{35}$$

Transforming Eqn (35) for the reliability of the system is written:

$$R \approx \prod_{i=1}^{m} \Phi(\beta_i) \tag{36}$$

where for simplicity of notation the symbols for the conditional probability are missing.

For simplification, the following assumptions are made. All members that are considered to have a potential crack are checked by a visual inspection method and if a crack damage is found in an element then it is replaced by a perfect one. The progressive fatigue failure of the structural system is considered. If n denotes the number of elements that have a repair at time T_j, and m is the total number of elements, then the reliability of the structure can be expressed as:

$$R_{j+1}(t) = \prod_{l=1}^{n} R^{r}_{l,j+1}(t) \prod_{k=n+1}^{m} R^{nr}_{k,j+1}(t) \tag{37}$$

where $t \in [T_j, T_{j+1}]$ and no correlation is considered.

DESIGN FOR MINIMUM REPAIR

One of the main advantages of first principles based probabilistic structural design is that the reliability of the ship structure with respect to the various possible modes of failure may be examined and quantified. Fatigue is one of the main time dependent factors affecting structural condition, which is in fact the main object here.

Reliability requirements may vary greatly depending of different parameters, describing the loading condition and structure itself. They may sometimes be set by the designer or more broadly by the Classification Societies. In other situations, the requirements may be imposed by the owner of the ship and in some instances, third parties such as government agencies may play a large role. For any structure there are likely to be trade-offs between minimum reliability level and repair works that have to be done, initial crack size which is related with quality of manufacturing of the structure, time between inspections, detectable crack size etc.

The approach presented here may be used as a decision tool for different reliability based maintenance policies. One is if the interval between inspections is known to be $\Delta t_{j+1} = \Delta t_o$ and the detection limit of the method of inspection is $a_{d,j+1} = a_d$, for $j \in [0,n]$ and n is total number of inspections. Then the reliability can be calculated:

$$R_{j+1}(t) = g_1(\Delta t_o, a_d) \text{ where } t \in [T_j, T_{j+1}] \tag{38}$$

The second case is when there is a minimum acceptable value of the reliability level $R_{\min}$ and the detectable crack size $a_{d,j+1} = a_d$, for $j \in [0,n]$, is fixed. Then the time interval between each inspection Δt_{j+1} can be calculated:

$$\Delta t_{j+1} = g_2[R, a_d] \text{ if } \{[R_{j+1}(t) \le R_{\min}] \cap [a_{d,j+1} = a_d]\}, \text{ where } \Delta t_{j+1} \in [\Delta t_{\min}, \Delta t_{\max}] \tag{39}$$

The third possible application is fixing the time interval between inspections, $\Delta t_{j+1} = \Delta t_o$ and the minimum level of the reliability $R_{\min}$, the calculated limit for the detectable crack size is:

$$a_{d,j+1} = g_3(R, \Delta t), \text{if } \{[R \ge R_{\min}] \cap [\Delta t = \Delta t_o]\}, \text{ where } a_{d,j+1} \in [a_{d,\min}, a_{d,\max}] \tag{40}$$

This implies the choice of the method of inspection that is able to accomplish it.

Another factor, which can also affect fatigue and the relative rate of replaced elements (proportion between the number of replaced elements in a specific location to the total number of replaced elements (*Nr*)), is the distribution on the net section area of stiffeners. It is known that the same inertia moment and modulus of the net section can be obtained with different location and net sectional area of the longitudinal girders. A differential expression, which is well known in structural design of ship structure, can be applied to determine the change of the midship section modulus $\Delta W(z)$ if some area $\Delta A(z_i)$ is added or removed in a specific place at the net-section z_i. The expression is written as:

$$\frac{\Delta W(z)}{W(z)} = \sum_i \frac{\Delta A(z_i)}{A_o}\left(\frac{(z_{NL} - z_i)^2}{\rho^2} + \frac{(z_{NL} - z_i)}{z_{NL} - z}\right) \tag{41}$$

where A_o is the total net-section area of the midship section, $\rho = \sqrt{I_y/A_o}$ is the radius of inertia, I_y is the moment of inertia with respect to the axis y and z_{NL} is the distance between the base line and the neutral line. Eqn (41) is transformed for the case in which the stiffener areas are redistributed with the restriction that the moment of inertia and the midship section modulus are kept constant, which leads to $\Delta W(z) = 0$. The transformed expression is given as:

$$\sum_i \Delta A(z_i)\left(\frac{(z_{NL} - z_i)^2}{\rho^2} + \frac{(z_{NL} - z_i)}{z_{NL} - z}\right) = 0 \tag{42}$$

Applying Eqn (42) different fatigue design problems can be solved, by optimising the location of the fatigue damage rate leading to a minimum of the relative rate of replaced elements.

NUMERICAL EXAMPLE

The formulation presented here is applied for the determination of the fatigue life of the side shell structure of a tanker with a deadweight of 398,800 *tons*, a length of 340 *m*, breadth of 56 *m* draft of 22 *m*, depth of 32 *m* and a block coefficient of 0.84. The yield stress was considered to be 390 *MPa* and the ship speed is 16 knots. The space between the stiffeners is b_s=1m and the space between transverse web frames is l_s=4.9m.

The moment of inertia corresponding to the axis y is I_y=1251 m^4 and to the axis z is I_z=2884 m^4 and the distance from the base line to the neutral axis in direction z is z_n=14.4 m and in the y to direction is y_n=0. The structural detail that is subjected to the reliability analysis is located on the side shell of ship and has co-ordinates $x = 0$ m, $y = 28$ m, $z \in [0,32]$ m.

The ship is required to operate with full central tanks and empty side tanks. In this situation, the influence of the internal pressure is neglected, but in a general case, this situation is more dangerous than in the case when the side tanks are full and the central tank is empty. The equations for the vertical and the horizontal bending moments and the wave pressure on the structure are taken from the rules of Classification Societies corresponding to the 10^{-8} probability level.

The potential cracks are considered on the intersections of the longitudinal stiffener with the transverse frames and bulkheads. The material constants are taken as C=1.7$10^{-11}$ and m=3. The initial crack size is $a_0 = 0.5$ mm and the critical size a_{cr} is equal of the height of the stiffeners. It is assumed, as suggested in [20], that the time to crack initiation is ten percent from the time of crack propagation until the critical size. The parameters of the Weibull distribution of time to crack initiation are taken as $\alpha_t = 2, \beta_t = 20$ and the geometry parameter is Y=1.12 which implies the same manufacturing conditions for all elements.

The reliability assessment considers that during the time for inspection all elements are observed. If the crack is detected in an element, then it is perfectly repaired. The basic results are produced by $a_d = 0.02\ m,\ \Delta t_o = 4$ years.

The approach presented may be used as a decision tool for different reliability based maintenance policies, some of which have been discussed in [7, 21]. This implies the choice of the method of inspection that is able to accomplish it. Depending of the inspection policies the reliability varies and it affects to the number of repaired elements and the minimum reliability (see Figure 1). The location of the maximum of the relative number of replaced elements can be varied applying different policies of detection, probability level of detection, time interval between inspections as has been demonstrated in [7]. As was already recognised, the maximum of the relative rate of replaced elements depends from many factors including environment, fabrication quality, loading, etc. It can be seen in Figure 1 that the maximum and distribution of the relative rate of replaced elements can be controlled by adopting different policies of inspection.

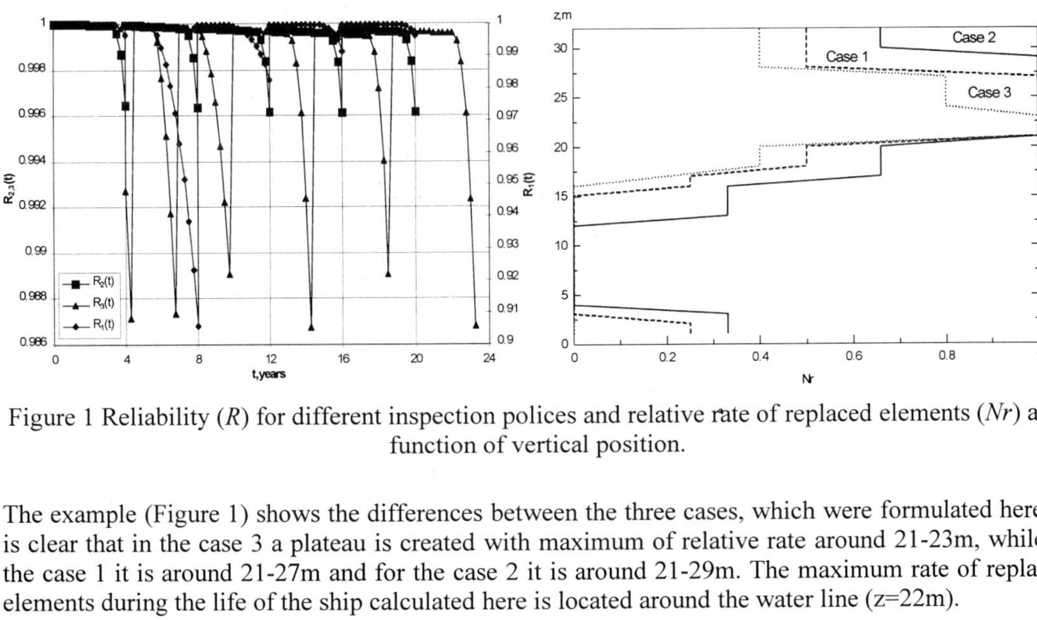

Figure 1 Reliability (R) for different inspection polices and relative rate of replaced elements (Nr) as a function of vertical position.

The example (Figure 1) shows the differences between the three cases, which were formulated here. It is clear that in the case 3 a plateau is created with maximum of relative rate around 21-23m, while in the case 1 it is around 21-27m and for the case 2 it is around 21-29m. The maximum rate of replaced elements during the life of the ship calculated here is located around the water line (z=22m).

Figure 2 shows the result of the redesign process, i.e., the change of the modulus of the stiffeners taking into consideration that the moment of inertia and the midship section modulus are the same as before transformation. The stiffener modulus which are located at z=21, 22, 23 and 24 m are increased from 351 cm^3 to 497 cm^3. This was compensated with increasing on the modulus of the stiffeners that are located at z=1, 2, 3 and 4 m from 2500 cm^2 to 2924 cm^2. The transformation was made applying Eqn (42), which leads to:

$$dW_s(z) = \frac{W_{s,\mathrm{mod}}(z) - W_{s,case\ 1}(z)}{W_{s,\mathrm{mod}}(z)} 100,\ [\%] \tag{43}$$

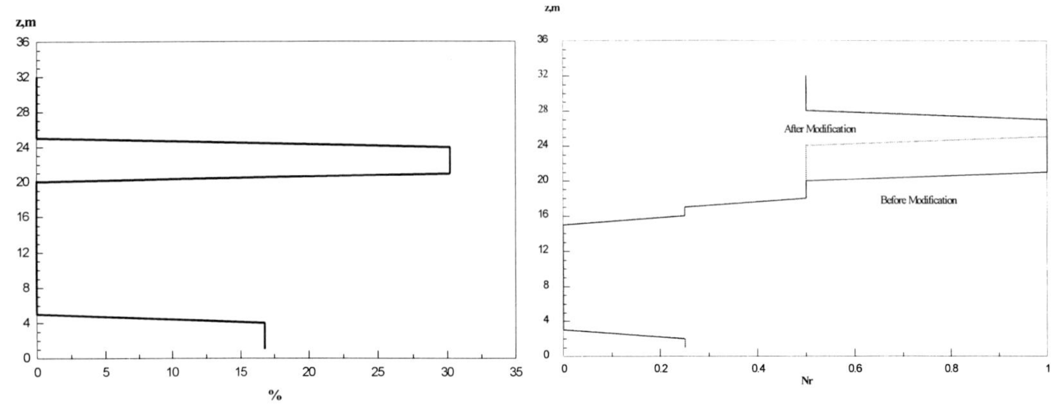

Figure 2 Change of the stiffener modulus and relative number of replaced elements (*Nr*) as a function of vertical position.

To examine how the mean wave response period, initial crack size, detectable crack size and time between inspections affects the reliability and the number of replaced elements, nine cases are studied as shown in Table 1. The variation of the parameters has been chosen because the intervals between inspection (and repair) are fixed for ships. This means that sometimes more critical situations may not result in a lower reliability. In fact, it may only indicate that the repair operations are made earlier.

TABLE 1 VARIATION OF PARAMETERS, [YEARS, MM]

	T, sec			a_o,mm			a_d, mm			*dt*,years		
	8	10	12	0.1	0.6	1.1	10	20	30	4	5	6
Case 1		•			•			•		•		
Case 2	•				•			•		•		
Case 3			•		•			•		•		
Case4		•				•		•		•		
Case5		•		•				•		•		
Case6		•			•		•			•		
Case7		•			•				•	•		
Case8		•			•			•			•	
Case 9		•			•			•				•

Figure 3 shows the reliability as a function of time for the various values of detectable crack size presented in Table 1. The right hand side of Figure 3 presents the polynomial approximations to the minimum reliability and relative number of replaced elements. Having a small value of detectable crack size gives opportunity for keeping reliability on a relatively higher level (Case 6). Having small a_d requires expensive techniques of inspection, which increases the repair work (number of detected and replaced elements) so as to increase the minimum reliability, as can be seen in Figure 3.

Initial crack size is related to the quality of manufacturing and it has great importance for fatigue reliability. As can be seen from Figure 4, increasing initial crack size leads to large reduction of minimum reliability, which is achieved during the ship life with a large number of repaired elements shown on the right side in the figure.

The average period of the sea state has a significant influence on the reliability, as shown in Figure 5. Its change will directly influence the number of cycles that the structure will be subjected to. The curve presented in right hand side of Figure 5 are derived by a regression approximation.

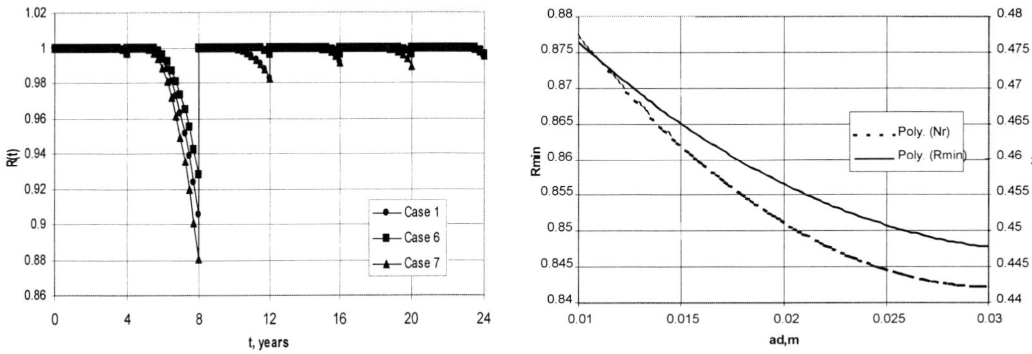

Figure 3 Reliability (R), relative number of replaced elements (Nr) and minimum reliability ($Rmin$) as a function of detectable crack size.

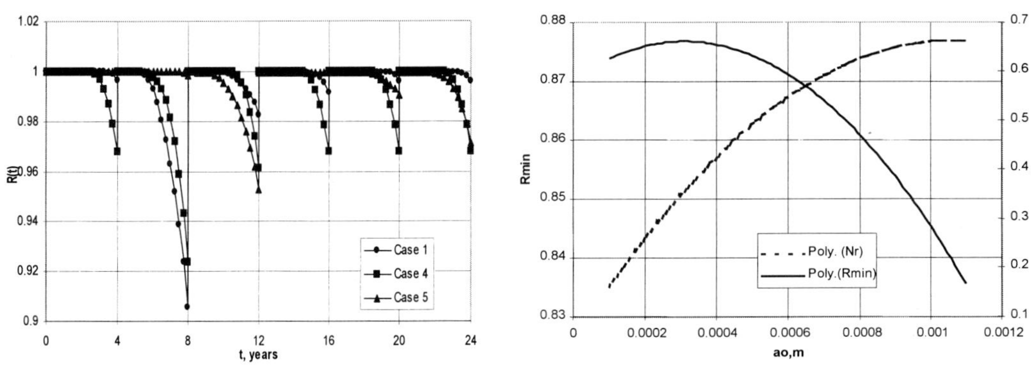

Figure 4 Reliability (R), relative number of replaced elements (Nr) and minimum reliability ($Rmin$) as a function of initial crack size.

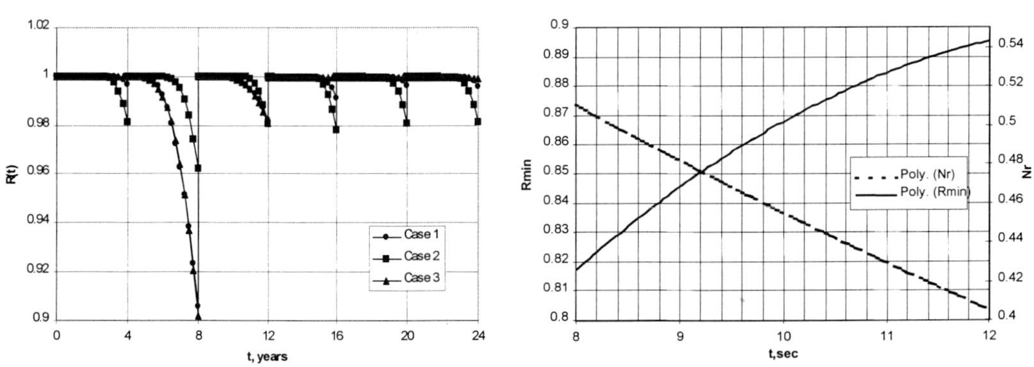

Figure 5 Reliability (R), relative number of replaced elements (Nr) and minimum reliability ($Rmin$) as a function of average period of the sea state.

Figure 6 shows the reliability as a function of time between inspections, which varies between 4 and 6 years. Decreasing the time interval will not allow the reliability to decrease as much as in the longer period. This can be seen also from Figure 6, which shows that when having a smaller interval between successive inspections a consistently higher minimum reliability level is achieved during ship operation. It is clear from the results that the time interval between inspection of approximately five years gives the minimum necessity of replaced elements.

Figure 7 shows that the minimum number of replaced elements may be achieved by keeping minimum reliability around 0.8- 0.85 which is also related with a certain time between inspections and with detected crack size. The mentioned value is valid only for the present example and the specific input data that have been used.

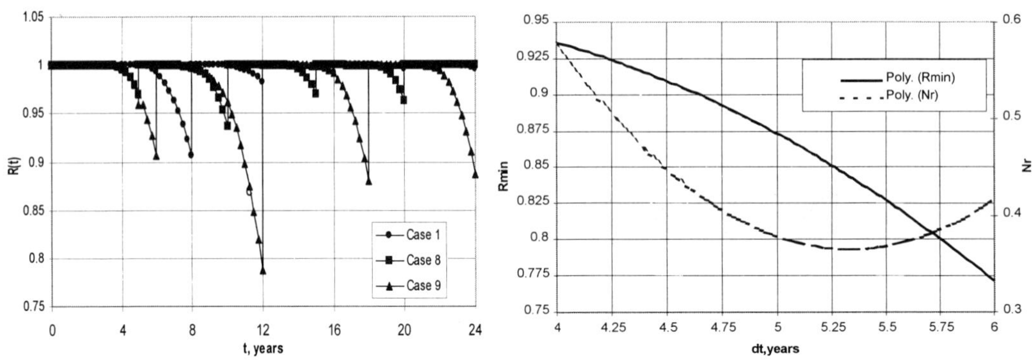

Figure 6 Reliability (*R*), relative number of replaced elements (*Nr*) and minimum reliability (*Rmin*) for various times between inspection.

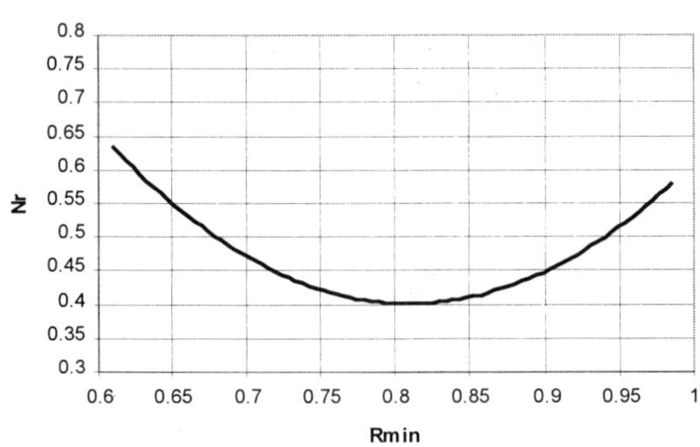

Figure 7 Relative number of replaced elements (*Nr*) as a function of minimum reliability (*Rmin*).

CONCLUSION

A formulation is presented for the assessment of the fatigue damage and the reliability of the side shell of the ship hull structure. The model accounts for the crack growth process applying linear elastic fracture mechanics. The long-term stress range acting on the elements is defined as a function of the local transverse pressure of internal cargo and outside water, combined with the stresses resulting from the longitudinal bending of the hull, including a combination of horizontal and vertical bending moments.

It is shown in the present paper that the relative number of repaired elements during the ship life varies as a function of the vertical position of the element in the side shell. It is demonstrated how this reliability based formulation can be used to redesign the side shell to achieve decreased repair costs.

It is also demonstrated how this formulation can be used for reliability based maintenance planning in particular how to vary the inspection interval in order to vary the maximum of repaired elements keep the same level of reliability. Alternatively, for fixed inspection intervals it is shown how the initial crack size, detectable crack size, time interval between inspection, average period of the sea state requirements of the minimum reliability and the number of repaired elements vary along the ship life. It is demonstrated that keeping a certain value of minimum reliability is reflected on the to minimum rate of replaced elements.

ACKNOWLEDGEMENTS

The work has been performed as part of the research project "Fatigue based design rules for the application of high tensile steels in ships (FatHTS)" which was partially financed by the European Union through the BRITE-EURAM programme under contract No. BE-95-1937, which involves the following additional participants: TNO, Registro Italiano Naval (IT), Bureau Veritas (FR), Chantiers de l` Atlantique (FR), Hamburg University (DE), Germanischer Lloyd (DE), Fincantieri (IT), Lisnave (PT), The Royal Schelde Shipyard (NL), Odense Steel Shipyard (DE), AF-Industriteknik (SE), Chalmers University of Technology (SE) and VTT (FI).

REFERENCES

1. Schall, G. and Ostergaard, C., 1991, Planning of Inspection and Repair for Ship Operation, *Proceedings of the Marine Structural Inspection, Maintenance, and Monitoring Symposium*, SNAME, pages VF1-VF7.
2. Guedes Soares, C. and Garbatov, Y., 1996, Fatigue Reliability of the Ship Hull Girder, *Marine Structures*, Vol. 9, No. 3, pp. 495-516.
3. Guedes Soares, C. and Garbatov, Y., 1996, Fatigue Reliability of the Ship Hull Girder Accounting for Inspection and Repair, *Reliability Engineering and System Safety*, Vol. 51, No. 2, pp. 341-351.
4. Cramer, E. H., Loseth, R., and Bitner Gregersen, E. M., 1993, Fatigue in Side Shell Longitudinal due to External Wave Pressure, *Proceedings of the 12th International Conference on Offshore Mechanics and Arctic Engineering (OMAE'93)*, Vol. II, ASME, New York, USA, Guedes Soares, C. et al. (Eds), pp. 267-272.
5. Cramer, E. H., Loseth, R., and Olaisen, K., 1995, Fatigue Assessment of Ship Structures, *Marine Structures*, Vol. 8, pp. 359-383.
6. Friis Hansen, P. and Winterstein, S. R., 1995, Fatigue Damage in the Side Shells of Ships, *Marine Structures*, Vol. 8, pp. 661-655.
7. Garbatov, Y. and Guedes Soares, C., 1997, Fatigue Reliability of Welded Joints in Tanker Structure, *Proceedings of the 16th International Conference on Offshore Mechanics and Arctic Engineering (OMAE`97)*, Vol. II, ASME, New York, USA, Guedes Soares, C. et al. (Eds), pp. 219-228.
8. Guedes Soares, C. and Moan, T., 1991, Model Uncertainty in the Long Term Distribution of Wave Induced Bending Moments for Fatigue Design of Ship Structures, *Marine Structures*, Vol. 4, pp. 295-315.

9. Det norske Veritas, 1993, *Rules for Classification of Ship, Hull Structural Design, Ship with Length 100 Meters and Above*, Det norske Veritas, Hovik, Norway.
10. Turkstra, C. J. and Madsen, H. O., 1980, Load Combinations in Codified Structural Design, *Journal of the Structural Division*, Vol. 106, ASCE, pp. 2527-2543.
11. Freudenthal, A. M. and Gumbel, G., 1954, Physical and Statistical Aspects of Fatigue, *Advances Applied Mechanics*, Vol. 4.
12. Corotis, R. B., Vanmarcke, E. H., and Cornell, C. A., 1972, First passage of Non-Stationary Random Processes, *Journal of the Engineering Mechanics Division*, Vol. 98, ASCE, pp. 401-414.
13. Packman, P. F., Pearson, H. S., Owens, J. S., and Young, G., 1969, Definition of Fatigue Cracks Through Non-destructive Testing, *Journal of Materials*, Vol. 4, No. 3, pp. 666-700.
14. Madsen, H. O., 1985, Random Fatigue Crack Growth and Inspection, *Proceedings of the 4th International Conference on Structural Safety and Reliability (ICOSSAR´85), Kobe, Japan.*
15. Rudlin, J. R. and Wolstenholme, L. C., 1992, Development of Statistical Probability of Detection Models using Actual Trial Inspection Data, *The British Journal of Non-destructive Testing*, Vol. 34, No. 12, pp. 56-64.
16. Delmar, M. V. and Sorensen, J. D., 1992, Probabilistic Analysis in Management Decision Making, *Proceedings of the 11th International Conference on Offshore Mechanics and Arctic Engineering (OMAE´92)*, Vol. II, ASME, New York, USA, Guedes Soares, C. et al. (Eds), pp. 273-282.
17. Lewis, E. E., 1987, *Introduction to Reliability Engineering*, John Wiley & Sons, Inc., New York, USA, pages 400.
18. Ditlevsen, O., 1979, Narrow Reliability Bounds for Structural Systems, *Journal of Structural Mechanics*, Vol. 7, No. 4, pp. 453-472.
19. Ditlevsen, O., 1979, Generalised Second Moment Reliability Index, *Journal of Structural Mechanics*, Vol. 7, pp. 435-451.
20. Bureau Veritas, 1984, *Cyclic Fatigue of Steel Ship Welded Joints*, Bureau Veritas, Paris.
21. Bea, R. G., 1994, Evaluation of Alternative Marine Structural Integrity Programs, *Marine Structures*, Vol. 7, pp. 77-90.

A METHOD FOR UNCERTAINTY QUANTIFICATION IN THE LIFE PREDICTION OF GAS TURBINE COMPONENTS

K. LODEBY
Volvo Aero Corporation, S-461 81 Trollhättan, Sweden
Mathematical Statistics, Chalmers University of Technology, S-412 96 Göteborg, Sweden

O. ISAKSSON
Volvo Aero Corporation, S-461 81 Trollhättan Sweden
Division of Computer Aided Design, Luleå University of Technology, S-971 87 Luleå,Sweden

N. JÄRVSTRÅT
Volvo Aero Corporation, S-461 81 Trollhättan, Sweden

ABSTRACT

A failure in an aircraft jet engine can have severe consequences which cannot be accepted and high requirements are therefore raised on engine reliability. Consequently, assessment of the reliability of life predictions used in design and maintenance are important.

To assess the validity of the predicted life a method to quantify the contribution to the total uncertainty in the life prediction from different uncertainty sources is developed. The method is a structured approach for uncertainty quantification that uses a generic description of the life prediction process. It is based on an approximate error propagation theory combined with a unified treatment of random and systematic errors. The result is an approximate statistical distribution for the predicted life. The method is applied on life predictions for three different jet engine components.

The total uncertainty became of reasonable order of magnitude and a good qualitative picture of the distribution of the uncertainty contribution from the different sources was obtained. The relative importance of the uncertainty sources differs between the three components. It is also highly dependent on the methods and assumptions used in the life prediction. Advantages and disadvantages of this method is discussed.

KEYWORDS

Uncertainty, Propagation, Life prediction, Gas turbine, Low cycle fatigue, Creep

INTRODUCTION

Volvo Aero Corporation has a responsibility over the entire life cycle for a number of jet engines and engine components. This requires continuos efforts to monitor and predict the expected life of the engines and their components, as well as improving the products by re-design and development efforts. Since a mechanical failure in these products can have severe consequences, which cannot be accepted, high requirements are raised on engine reliability. Consequently, the methods used in prediction of expected component life must be assessable.

The life prediction process

Where no in-service experience exists, analytical life prediction methods have to be used. Often, these are based on deterministic predictions resulting in a single value. In generic terms, the life prediction process description is illustrated in Figure 1. This process consists of a series of linked activities, which requires specific methods and data depending on the situation. The degree of idealisation in the simulation model varies depending on the conditions for the simulation [11]. In each of these activities, uncertainties are introduced due to the scatter in the data and approximations in the methods used.

The generic activities in the middle uses methods (to the right) and data (to the left). Which methods and data that are used depends on the situation. The required accuracy together with available resources and data are the most important parameters deciding which approach to chose.

The *problem* set up has to be defined and an analysis strategy has to be outlined. A *geometry model* is needed as a base for the *computational model* which defines the conditions (loads, topology and boundary conditions). *Thermal analyses* are made since the thermal field is vital for many engine components since high temperature and temperature gradients causes thermal fatigue [2], and material properties may be highly temperature dependent. The *structural analyses*, which are based on the structural loads and thermal fields then results in information of the stress/strain field. Once the temperatures and stresses/strains are known, various *life prediction* models can be applied to predict the component life. Different models are used depending on the situation, e.g. crack propagation models, crack initiation models and oxidation models. Finally, the predicted life must be assessed using some reliability model in a *risk analysis*. Uncertainties in the data used, assessments made and accuracy of methods used have to be taken into account, which is being addressed in this work. For analysis of components where field experience exists, these data can be used to perform the risk analysis.

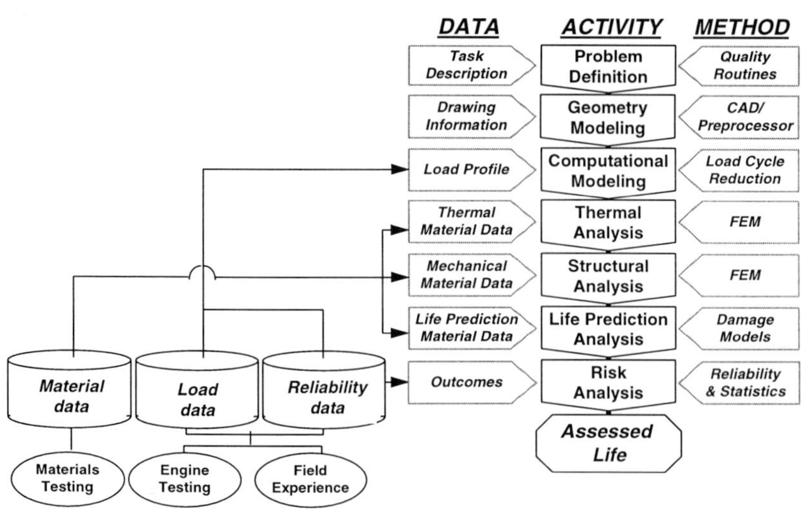

Figure 1: Generic description of the life prediction process at Volvo Aero Corporation, from [4].

Basically, uncertainties can be compensated for using conservative values in each activity in the process and obtain a minimal life. As an example, 1/1000 risk can be used for Low Cycle Fatigue (LCF) data as illustrated in Figure 2. This approach is undesirable since for most situations this lead to an oversized component, or an underestimated life.

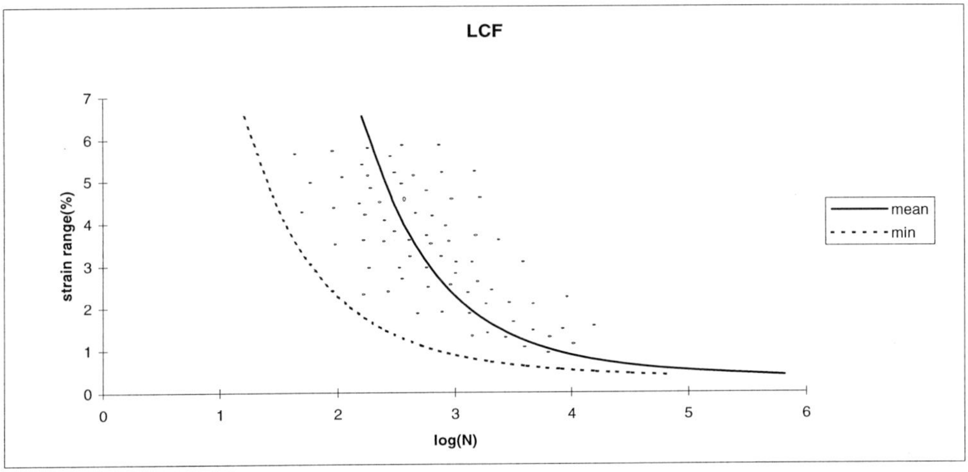

Figure 2: Strain range vs. life to crack initiation. The mean- and minimum-curves plotted in a statistically distributed data scatter.

Preferably, a probabilistic life prediction model, including the variability and uncertainty of the parameters, should be used. A probabilistic life prediction model results in a cumulative distribution function, for the estimated life rather than a single deterministically predicted value. Examples of probabilistic failure models can be found in [7] and [9].

In this work, the uncertainties introduced in each activity are identified, categorised, quantified and finally summed to obtain the total uncertainty of the entire simulation. This approach gives an idea of the effect from individual uncertainty sources on the predicted life, and critical parameters in methods and data can be identified.

The method has been applied on three different life predictions of jet engine components.

METHOD

The method described in this paper is derived from an American national standard (ASTM) for measurement uncertainties [3]. This standard specifies procedures for the evaluation of individual uncertainties, arising from random and systematic errors, and for the propagation of these errors into the final result. A similar methodology has been used for high cycle fatigue with variable amplitude in [8].

The activities considered are the four analysis activities in Figure 3. The figure illustrates the principle of uncertainty analysis in the situation where no field experience is available.

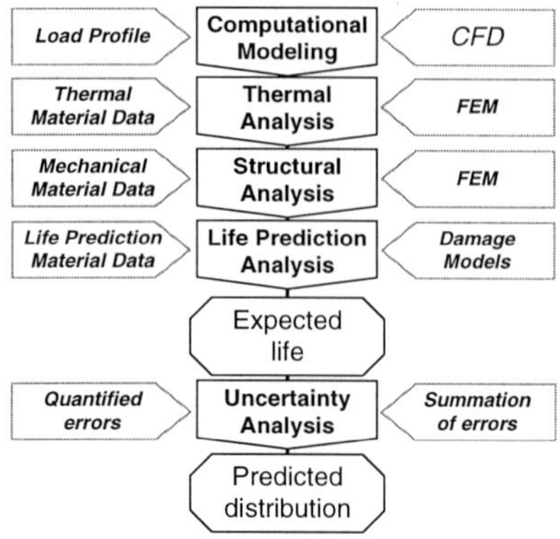

Figure 3. Uncertainty analysis applied on the life prediction process

A deterministically predicted life is obtained from the analytic life prediction and after the uncertainty analysis, an approximate statistical distribution for the predicted life is obtained. The methodology can be described in six main steps briefly described below.

Identify the uncertainty sources

The sources of variability and uncertainty have to be identified. In each activity the random nature of the input data and the approximations in the methods have to be examined, see Figure 4.

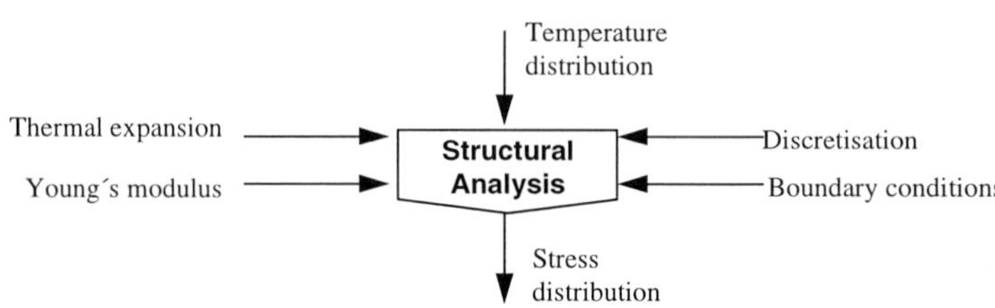

Figure 4. Inputs associated with uncertainties in an activity

Categorize into random and systematic errors

The uncertainties are either random or systematic. The random errors are due to scatter in the input data, e.g. the Young's modulus of the material. In our case, the systematic errors, or biases, are uncertainties due to approximations and idealisations in the methods used.

Quantification of random errors

The random errors are assumed to be normally distributed and are quantified in terms of the standard deviation. The scatter can usually be determined from measured data, but sometimes estimates based on previous experience must be used.

Quantification of systematic errors

The systematic errors (the biases), are constant in repeated measurements or calculations. In order to determine a bias exactly, it would be necessary to compare with the true value, which is usually impossible. However, it is often possible to estimate bounds for the error. Here, these errors are considered symmetric.

For the life predictions considered the systematic errors are due to idealisations [11]. The following methods for estimation are used. In order of preference:

1. Comparison of the calculated value with measurements.
2. Comparison of two methods of different accuracy
3. Estimation based on experience

The systematic errors are deterministic deviations, but in practise the magnitudes are unknown. Within the limited scope of this work the systematic errors are considered to be normally distributed random errors. The reason for this assumption is that we will allow an error of any size but the variance of the error is bounded. The most general distribution with bounded variance is the normal distribution in the sense that it maximises the entropy given the restraint of bounded variance. Using estimated maximum errors for the 95 percentile point, an approximate standard deviation is obtained by dividing this estimate by 2. Still, the errors are categorized in random and systematic errors since we are interested in separating the contribution from the two categories.

Examine the correlation between the uncertainty sources

The uncertainty sources may have a coupled effect on the predicted life. This effect have to be accounted for and is later introduced as a correlation.

Sum all uncertainties by Gauss Approximation formula

As the uncertainties are quantified for all activities their contributions are summed using the Gauss Approximation formula [6].

As an example, suppose that X_i are random variables representing the basic parameters with quantified uncertainties. That is, estimated values of the variances are available and the variance of Y is needed. If Y is related to X_i as $Y = f(X_1,X_2,...X_n)$ and, the variance of Y is given by:

$$\sigma_Y{}^2 \approx \sum_i^n \sum_j^n \rho_{ij} \sigma_{X_i} \sigma_{X_j} (\frac{\partial f}{\partial X_i \partial X_j}) = \sum_i^n (\sigma_{X_i} \frac{\partial f}{\partial X_i})^2 + \sum_i^n \sum_{j \neq i}^n \rho_{ij} \sigma_{X_i} \sigma_{X_j} (\frac{\partial f}{\partial X_i \partial X_j}) \qquad (1)$$

where ρ_{ij} the correlation of X_i and X_j. The derivatives are calculated in the mean of the distribution of X_i. Since even the systematic errors are treated as random variables, (1) can be used for all errors. This gives an approximate standard deviation for the predicted life as well as quantified contributions to this standard deviation from each uncertainty source.

EXAMPLES

To illustrate the principle of the method, three case examples have been studied.

1. The combustor wall of a stationary gas turbine.
2. The flame holder of a military jet engine.
3. A high pressure turbine-blade of a military jet engine.

Each of these life prediction analyses has been performed at Volvo Aero Corporation under different conditions in terms of lead time and required accuracy.

In each of the analysis activities, a number of uncertainties are considered. Throughout the paper, only a critical point or area is considered, i.e. the point where failure is expected to occur according to the life prediction.

The combustor wall in a stationary gas turbine

The combustor wall is exposed to almost pure thermal load. Since the load in service is held almost constant, with long operating hours at high temperatures, the failure mode is creep.

Computational modelling. In this case this activity involves calculation of thermal boundary conditions using computational fluid dynamics (CFD), which predicts the fluid velocities and the gas temperatures along the combustor wall. The CFD analysis is followed by an energy balance calculation, in order to determine the heat transfer coefficients. The load is approximated to be constant at maximum, which is a good assumption in this case. The important output parameter from this activity is considered to be the heat flow, Q_{tot}, in the critical area.

The uncertainties in the CFD-analysis are due to uncertainty in performance data, geometry simplifications and discretisation. This causes errors in predicted fluid velocities and gas temperatures. The magnitudes have been estimated by specialists.

In the following heat balance calculation additional uncertainties are introduced as circumferencial variations are neglected and one dimensional heat conduction is assumed. An upper limit of the magnitude of these errors is estimated.

Thermal analysis. The uncertainties introduced in the thermal analysis are scatter in the thermal conductivity and systematic errors due to geometry simplification and discretisation.

Structural analysis. The uncertainties introduced in the structural analysis are scatter in the Young's modulus and the thermal coefficient of expansion. Systematic uncertainties are introduced due to discretisation, uncertainties in the mechanical boundary conditions and in the reduction of the of the multiaxial stress condition.

Life prediction. The time to creep failure is predicted using the Larson-Miller parameter [8]. The uncertainties introduced in this final prediction are scatter in the Larson-Miller parameter and systematic errors due to the use of this simple creep model and extrapolations. Also, creep may be the dominating failure mode, but neglecting other failure modes introduces uncertainty.

Propagation. The sources of uncertainty are identified and the magnitudes have been estimated by specialists. In order to sum the uncertainties the relation between the input parameters and the life are needed, i.e. the function, f, in (1) is needed. This function is quite complicated to derive analytically. However, only the derivatives are needed. In order to find the derivatives the serial structure of the life prediction is used. For example the derivative of the time to failure with respect to fluid velocity, u, is found by:

$$\frac{\partial \log(t)}{\partial u} = \frac{\partial \log(t)}{\partial \sigma} \frac{\partial \sigma}{\partial T} \frac{\partial T}{\partial Q_{tot}} \frac{\partial Q_{tot}}{\partial u} \tag{2}$$

where t is the time to failure, σ is the stress in the critical point and T, the corresponding, temperature. Q_{tot} is the total heat flow through the wall in the interesting region. The logarithm is used since the life is known to be approximately lognormally distributed.

The factor $\frac{\partial \log t}{\partial \sigma}$ is derived from the Larson-Miller diagram. To find $\frac{\partial \sigma}{\partial T}$, the relation between the temperature and the temperature gradient in the critical point is needed. This has been found by pure reasoning. The relation between the temperature gradient and the stress is the coefficient of thermal expansion. $\frac{\partial T}{\partial Q_{tot}}$ and $\frac{\partial Q_{tot}}{\partial u}$ are found by setting up an energy balance equation for the heat flow through the wall and differentiate.

Correlation. For simplicity the correlation of the errors are assumed to be either 0 (no correlation)or 1 (full correlation). The error in temperature enters the life prediction in two ways. First, because the creep damage rate is temperature dependent and second, due to the thermal expansion, which causes a change in the stress. These contributions are clearly dependent, since they originate from the same source, and therefore $\rho=1$ is assumed. In all other cases $\rho=0$ are assumed.

Summation. Equation (1) yields the total uncertainty in the prediction in terms of the standard deviation for log(t). Since the total uncertainty is expressed as a sum it also expresses the contributions from each uncertainty source.

Flame holder in a military jet engine

The flame holder is located after the turbine, in the afterburner and its function is to create stable combustion in the afterburner. It is exposed to almost pure thermal loading and the failure mode is considered low cycle thermal fatigue [2].

The uncertainty analysis is carried out in the same manner as for the combustor. The main differences are the different failure mode and that the thermal boundary conditions were obtained by measurements. Furthermore, the stresses are above the yield strength for the material and a compensation for this is done by the Neuber-rule [5], since the structural analysis is linear. The life is predicted using the Coffin-Manson-Morrow equation [1].

The consequences for the uncertainty analysis are that instead of studying the uncertainties in the CFD-analysis the measurement uncertainties has to be considered. In the final life prediction, the magnitude of the error due to the plasticity is estimated by a comparison with a linear rule. No correlation between the uncertainty sources were taken into consideration.

A high pressure turbine blade in a military jet engine

The turbine blade is exposed to thermal, rotational and pressure load and the failure mode considered is low cycle fatigue. The uncertainty analysis is carried out in exactly the same manner as above. No correlation is assumed.

RESULTS

In absolute terms the total uncertainty became of reasonable order of magnitude. The figures below shows the relative uncertainty contribution from the different activities divided in random and systematic uncertainties, in percent of the total variance.

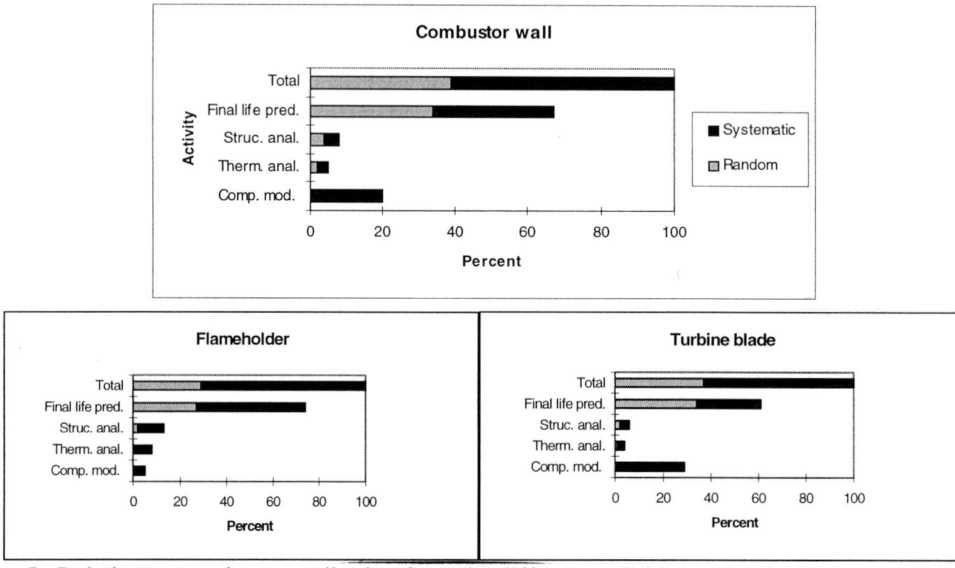

Figure 5. Relative uncertainty contribution from the different activities in the three cases divided in random and systematic uncertainties.

It can be seen that the prediction uncertainty is dominated by the contribution from uncertainties in the final life prediction. The random part of this comes from scatter in the life prediction material data and is the largest single uncertainty source in all three cases. The contributions from the finite element analyses are small. In the combustor analysis, the time to failure at maximum load is computed and the uncertainty in the load cycle reduction is therefore not considered. Thus, the contribution from the computational modelling is solely due to the uncertainties in the computation of the thermal boundary conditions.

In the flameholder analysis the thermal boundary conditions are measured and the uncertainty in the measurements are attributed to the thermal analysis. The uncertainty contribution from the computational modelling is therefore, (as opposed to the combustor analysis) solely due to the uncertainty in load cycle reduction.

DISCUSSION

Regarding the presented results, several questions can be raised about the reliability of the uncertainty evaluation. Scatter in life prediction material data gives the largest contribution to the total uncertainty. This might be true, but the deviation between the real load situation and the approximated load situation has not been assessed in detail. Consequently, the predicted life is related to the simplified load history, not the real situation. This type of uncertainties are not thoroughly treated in this work, and should be the topic for further work.

CONCLUSIONS AND FURTHER WORK

Several advantages and disadvantages can be identified using this method, see Table 1.

Table 1. Advantages and disadvantages using the presented method

Advantages	Disadvantages
Evaluation of the total uncertainty in the prediction is based on the sources of uncertainty.	Quantification of the elementary uncertainty sources is in some cases very difficult and may be extremely overconservative or even unconservative.
Results in an approximate distribution instead of a single conservative value.	Often, failures occur due to unforeseen failure modes, and this uncertainty cannot be included.
Highlights the problem of uncertainty.	

Ideally, this uncertainty evaluation method can be used to evaluate the reliability of analytical life predictions. At present the error estimates are to uncertain for the method to be used with confidence in design work. However, as for most structured methods, the major benefit initially is the accompanying discussion and communication between specialists and others interested in the result. The importance of this effect cannot be neglected. More work is required to ensure that the method becomes stable and reliable. Furthermore, a better way to represent different kinds of model errors and experience in the different error estimations are needed.

Finally, although there exists several difficulties and limitations it is a way to systematically address the issue of uncertainty evaluation of analytical life predictions. The underlying statistical treatment is straightforward, which justifies further work on the method.

The long time goal is to use a probabilistic model to incorporate the design parameters variability and uncertainty to obtain a cumulative distribution function for the life, instead of a conservative value, associated with an unknown risk level.

ACKNOWLEDGEMENT

Life prediction calculations at Volvo Aero by Martin Öman, Joakim Berglund and Ken Spiers were used in the examples. Part of the work was financed by NFFP (Nationella flygtekniska forskningsprogrammet) project no NFFP347, and the ENDREA programme.

REFERENCES

1. Bannantine, Fundamentals of Metal Fatigue Analysis, Prentice Hall, New Jersey, United States, 1990
2. Halford, *Low Cycle Thermal Fatigue*, NASA memorandum, 87225, 1986
3. Instruments and Apparatus supplement, ANSI/ASME PTC 19.1-1985, *Part 1: Measurements uncertainty*

4. Isaksson O, *Engineering Design Systems Supporting Integrated Product Development,* Licentiate thesis, Department of Mechanical Engineering, Luleå University of Technology, Luleå, 1997

5. Neuber H, *Theory of stress concentration for shear-strain prismatic bodies with arbritrary nonlinear stress-strain law*, Journal of applied mechanics, pp 544-550, dec 1961

6. Rice J.A, *Mathematical statistics and data analysis*, International Thomson Publishing, 1995

7. Rapp D, *Reliability assessment of thrust chamber cooling concepts using probabilistic analysis techniques,* Sverdrup Technology Inc, Lewis Research Center Group, Brook Park, Ohio, USA, 1993

8. Svensson T, *Prediction uncertainties at variable amplitude fatigue*, International Journal of Fatigue, 1997, **19**(1), 295-302

9. Tryon R & Cruse T, *Failure Model Development for an Integrally Bladed Turbine Wheel,* Vanderbilt University, Nashville, TN, USA, 1992

10. Webster G.A, Ainsworth, R.A, *High temperature component life assessment*, Chapman & Hall, 1994.

11. Wentorf, R., Shepard, M.S., *Automated Analysis Idealisation Control*, Concurrent Engineering: Automation, Tools, and Techniques, Ed. A. Kuisak, pp 41-73, John Wiley & Sons, 1993

THE PROBABILITY OF SUCCESS USING DETERMINISTIC RELIABILITY

K. WALLIN
VTT Manufacturing Technology, P.O.Box 1704, FIN-02044 VTT, Finland

ABSTRACT

Many material properties are, even today, often treated as deterministic properties corresponding to a specific material and loading condition. In such cases the inherent statistical nature of properties is not at all accounted for in design. Design criteria for fatigue represent a higher statistical awareness as they are usually intended to correspond to a specific statistically defined confidence level, like 2 or 3 standard deviations below the mean. However, even a simple S-N data set allows the use of several different analysis methods (some more proper than others), and the resulting design criteria are not always unambiguous. This is especially the case when the data set includes non-failed tests. In this work a maximum likelihood expression allowing for random censoring is presented and discussed in comparison to more often used methods of least squares.

KEYWORDS

Statistical analysis, rank probability, Weibull, Lognormal, random censoring, S-N data.

INTRODUCTION

Historically material properties have usually been treated as deterministic properties corresponding to a specific material and loading condition. Scatter in test results has often been attributed to errors in the test performance or to macroscopic material inhomogeneities. This "deterministic" thinking is reflected in design criteria. Often, the inherent statistical nature of properties is not taken into account during design. A typical example of such a deterministic thinking, is the ASME fracture toughness reference curve (K_{IR}) which is assumed to constitute an absolute lower bound for fracture toughness. In reality, this is not the case. It represents only a deterministic lower bound curve to a specific set of data, which represent a certain probability range. In the case of fatigue, there appears to have been increased awareness of the statistical aspects of the event. Design criteria for fatigue are usually intend to correspond to a specific statistically defined confidence level, like 2 or 3 standard deviations below the mean. However, several different analysis methods (some more proper than others) can be applied to even a simple S-N data set and the resulting design criteria are not always unambiguous. This is especially the case when the data set includes non-failed tests. This paper will highlight and discuss some possible statistical evaluation methods for S-N data, including non-failed tests, using an exemplary data set. Specifically, the difference between maximum likelihood and least square methods is addressed.

DATA SET AND DETERMINISTIC ANALYSIS

The data set used for the analysis were taken from a work by Marquis [1]. It consists of spectrum loaded tests of two different geometries: fillet weld specimens and box beam weld specimens. All details of the data set can be found in [1] and are not repeated here. The data is presented in Fig. 1. Noteworthy of the data is that three of the fillet weld results and the majority of the box beam weld results were non-failures.

Included in Fig. 1 is a "deterministic" lower bound curve drawn as an "eye-ball" fit through the lowest failure data. Interestingly, this deterministic lower bound is nearly identical to the LUT fillet weld design curve [1]. The deficiency of a lower bound curve of this type is that it does not give any direct quantifiable statistical information about the data. It is quite clear, that an estimate of this kind does not constitute a true deterministic lower bound. Instead, it corresponds to a specific probability range specific to the sample size. A true lower bound cannot be directly determined from test data, since it would require extrapolation to zero probability. The cumulative probability level represented by the lowest result can be expressed as [2]

$$P_{lowest} = 1 - (1 - P_{conf})^{1/n} \tag{1}$$

where P_{lowest} is the cumulative probability, P_{conf} is the confidence level and n is the sample size.

For a conservative estimate, P_{conf} should be taken larger than 50 %, eg. 80% or 95%. For the present data set the effective number of n is 85 (number of failed results and non-failed results larger than the lowest failure value). Thus with a 90 % confidence interval, 5 % to 95 %, the deterministic lower bound curve corresponds to a cumulative failure probability between 0.06 %...3.46 %, the median value being 0.8 %.

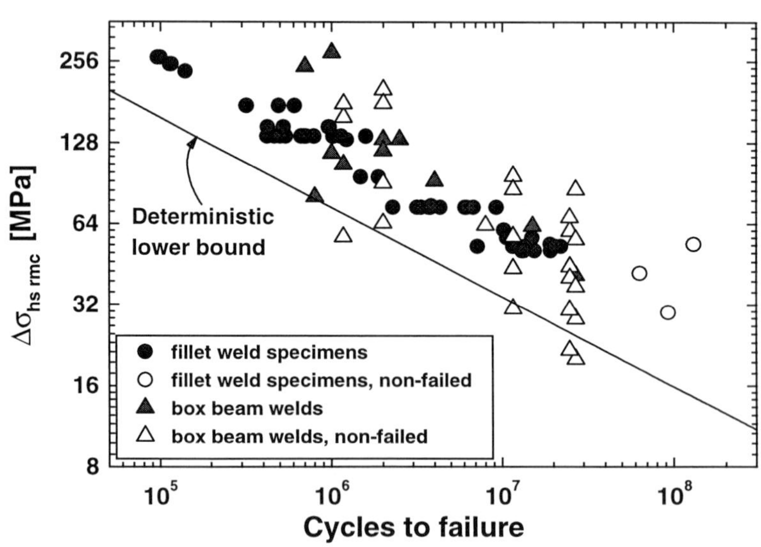

Fig. 1. S-N data set used for the statistical evaluation. Deterministic lower bound refers to an empirical "eyeball" fit through lowest failure point.

SIMPLE LEAST SQUARE FIT

The most common statistical method of analysing S-N data is the least square fit, LSF, method. Because of the nature of the least square method, only failed data can be used (Fig. 2). The method assumes implicitly that the data follows a constant Lognormal distribution. The resulting LSF for all failed specimens is presented in Fig. 3.

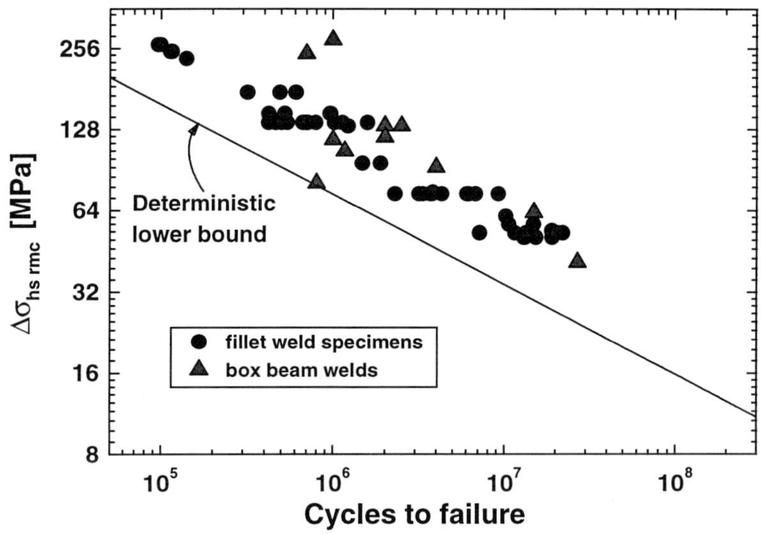

Fig. 2. S-N data set used for the statistical evaluation. Only failed specimens included. Deterministic lower bound refers to an empirical "eyeball" fit through lowest failure point.

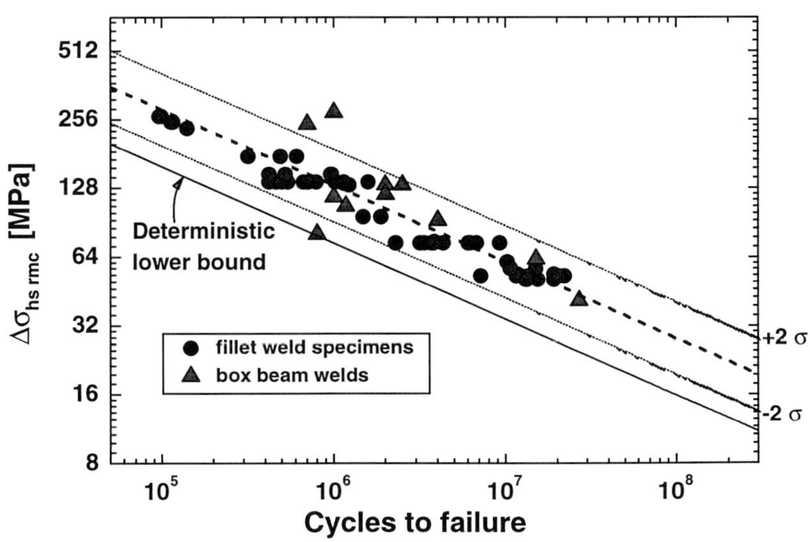

Fig. 3. LSF analysis of S-N data set. Only failed specimens included. Deterministic lower bound refers to an empirical "eyeball" fit through lowest failure point. $\pm 2\cdot\sigma$-lines refer to LSF.

The data in Fig. 3 is seen to be well described by a straight line in a log-log plot, which indicate a simple relation between stress amplitude and cycles to failure, i.e., $\Delta\sigma_{hs\ rmc}{}^{m} \cdot N$ = constant. The power m was fitted as being 3.1 for the combined data. Using a fixed power of 3 did not, however, significantly affect the goodness of fit. For both exponents, the standard deviation of stress was 22 %, which compared to the deterministic lower bound appears to be too small. The two sets of failed specimen data were also fitted individually to see whether the LSF analysis would indicate any significant differences between the two specimen geometries. The resulting mean fits are presented in Fig. 4 showing all specimens. The dashed line in Fig. 4 refers to the combined fit. The difference between the mean fits to the different geometries were not significant, but the scatter appeared significantly greater for the box beam welds. The LSF m was for the fillet welds 3.1, and for the box beam welds 2.7. Both values are, however, so close to 3 that a fixed slope describes the data just as well.

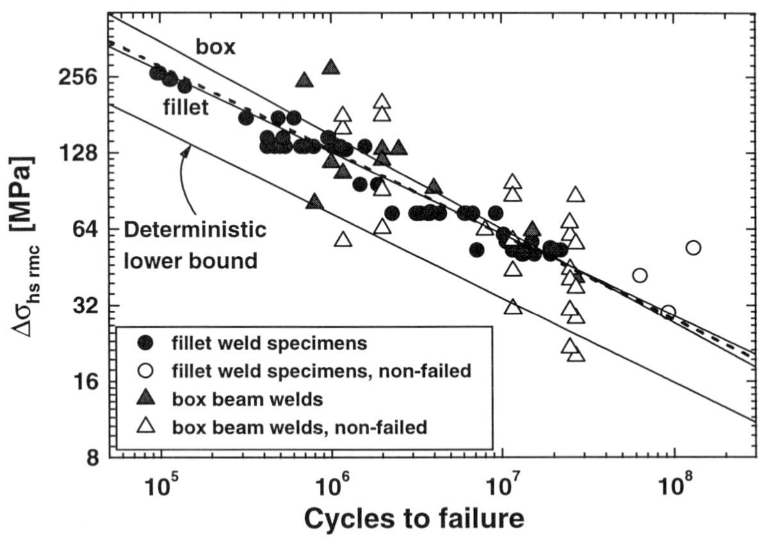

Fig. 4. LSF analysis of S-N data set. Only failed specimens included in LSF. Deterministic lower bound refers to an empirical "eyeball" fit through lowest failure point.

The problem with the LSF analysis is that it is not correct from a statistical point of view. By leaving out the non-failure results, the statistical relevance of the sample is distorted. The only way to properly analyse a data set including non-failed specimens, is to apply censoring to the data. This requires however the fitting of a specific distribution function, e.g., Weibull or Lognormal, to the data. A maximum likelihood estimation method, MML, allowing for random censoring is suitable.

MAXIMUM LIKELIHOOD ESTIMATION OF THE 2-PARAMETER WEIBULL EQUATION

As an example, the randomly censored MML estimate of the 2-parameter Weibull distribution is presented. The maximum likelihood method does not make use of the cumulative probability distribution. Instead it uses the probability density function directly. In this way, no information regarding the individual probabilities is needed. The probability density function for the 2-parameter Weibull distribution function is

$$f_c(x) = \frac{b \cdot x^{b-1}}{x_0^{\,b}} \cdot \exp\left(-\left\{\frac{x}{x_0}\right\}^b\right) \tag{2}$$

The MML method examines the likelihood that a certain probability density function describes a data set correctly. This is achieved by calculating a combined likelihood by multiplying all different discrete probabilities, i.e. the MML estimate is defined as

$$L = \prod_{i=1}^{n} f_c(x) \Rightarrow$$
$$L = \prod_{i=1}^{n} \frac{b x_i^{\,b-1}}{x_0^{\,b}} \cdot \exp\left(-\left\{\frac{x_i}{x_0}\right\}^4\right) \tag{3}$$

Eq. 3 gives the likelihood that a certain Weibull distribution describes the data. The parameters x_0 and b are solved as to produce a maximum in the likelihood L. This can be performed numerically using eq. 3 directly, or analytically, taking the derivative of eq. 3 with respect to x_0 and b.

The "normal" maximum likelihood method does not include censoring, but censoring is easily implemented by making use of the survival distribution function (S_c).

Any data in the set not corresponding to failure is not a part of the probability density function, but, rather, the survival distribution function. This leads to a conditional probability including both the probability density function as well as the survival function

$$\textit{Conditional probability} = f_c(K_c)^{\delta_i} \cdot S_c(K_c)^{1-\delta_i} \tag{4}$$

The parameter δ_i defines if the data point belongs to the probability density function ($\delta_i = 1$) or the survival function ($\delta_i = 0$).

The 2-parameter Weibull distribution survival function is

$$S_c(x) = \exp\left(-\left\{\frac{x}{x_0}\right\}^b\right) \tag{5}$$

and the censored maximum likelihood expression becomes thus

$$L = \prod_{i=1}^{n} f_c(x)^{\delta_i} \cdot S_c(x)^{1-\delta_i} \Rightarrow$$
$$L = \prod_{i=1}^{n} \frac{b \cdot (x_i)^{(b-1)\cdot\delta_i}}{(x_0)^{b\cdot\delta_i}} \cdot \exp\left(-\left\{\frac{x_i}{x_0}\right\}^b\right) \tag{6}$$

Equation 6 is solved with a method similar to that used for eq.3. The interesting feature of eq. 6, is that it does not restrict the censoring to any specific part of the data set, but each data point is treated individually. Thus, random censoring is actually implemented.

Solving eq. 6 is simplified by taking the logarithm of L (maximum of L is equivalent to maximum of ln[L]).

$$\ln L = \sum_{i=1}^{n} \left(\ln(b) + (b-1)\delta_i \cdot \ln(x_i) - b \cdot \delta_i \cdot \ln(x_0) - \left\{ \frac{x_i}{x_0} \right\}^b \right) \tag{7}$$

By solving for $\partial \ln(L)/\partial x_0 = 0$ and $\partial \ln(L)/\partial b = 0$ the randomly censored maximum likelihood estimate of the 2-parameter Weibull distribution is obtained as

$$\frac{\sum_{i=1}^{n}\delta_i}{b} + \sum_{i=1}^{n}\delta_i \cdot \ln(x_i) - \frac{\sum_{i=1}^{n}\delta_i \cdot \sum_{i=1}^{n}(x_i)^b \cdot \ln(x_i)}{\sum_{i=1}^{n}(x_i)^b} = 0$$
$$x_0 = \left(\frac{\sum_{i=1}^{n}(x_i)^b}{\sum_{i=1}^{n}\delta_i} \right)^{1/b} \tag{8}$$

In the case of S-N data where there are two interrelated parameters, $\Delta\sigma$ and N, one can specify a functional form for x_i like $x_i = \Delta\sigma_i^m \cdot N_i$ (or $x_i = (\Delta\sigma_i - \Delta\sigma_{th})^m \cdot N_i$ [3]), and to solve also m (and $\Delta\sigma_{th}$) by modifying the MML expression. For simplicity, in this study it was decided to fix the power m to 3, thus enabling the definition $x_i = \Delta\sigma_i^3 \cdot N_i$, i.e. S-N data is expressed in the form of a single parameter.

RANK PROBABILITY ANALYSIS

For graphical presentation an estimate of the individual probability for each result is needed. This is most easily achieved through a rank probability analysis. In a rank analysis, data are ordered by size and each data point is given a specific cumulative probability. The weakness with the rank probability estimates are that they are not measured values, but estimates of cumulative probability based on order statistics. Each data point corresponds to a certain cumulative failure probability with a certain confidence. This can be expressed in a mathematical form, using the binomial distribution, as

$$z' = 1 - \sum_{j=1}^{i} \frac{n!}{(j-1)! \cdot (n-j+1)!} \cdot P_{rank}^{j-1} \cdot (1 - P_{rank})^{n-j+1} \tag{9}$$

where z' is the probability that the rank estimate corresponds to the cumulative probability P_{rank}, n is the number of points and i is the rank number. Eq. 9 can be used to calculate rank confidence estimates. Simple approximations of the median rank probability estimate (z' = 0.5) are usually preferred. Three common estimates of the median rank probability are [4]

$$P_{rank} = \frac{i - 0.5}{n} \quad (10a)$$

$$P_{rank} = \frac{i}{n+1} \quad (10b)$$

$$P_{rank} = \frac{i - 0.3}{n + 0.4} \quad (10c)$$

The three approximations are compared with the outcome of Eq. 9 in Fig. 5. It seen that Eq. 10c is clearly the best estimate of the median rank probability. This definition of rank probability was used in the present analysis.

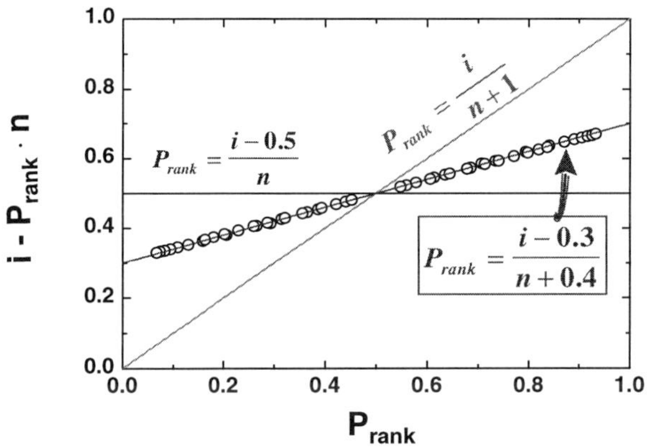

Fig. 5. Comparison of different estimates of median rank probability (lines) with binomial theory estimate (circles).

The use of the rank probability limits censoring to the upper end of the distribution, i.e., all censored values must be higher than any of the uncensored values. However, as in the simple LSF analysis of the basic data, often only failed data are included. This is not correct, but this censoring scheme is here applied to simplify the example.

The failed fillet weld specimens are presented in Fig. 6. The failed data does "apparently" follow a Weibull distribution fairly well. Figure 6 also includs a simple LSF to the rank probability data. This fit differs from the basic LSF presented earlier, in that here the data is fitted to a Weibull distribution, whereas a Lognormal distribution was implicitly assumed earlier. When the non-failed data are included in the analysis, the picture clearly changes (Fig. 7). If all three non-failed specimens would have represented the highest values, Fig. 7 would be correct from a statistical point of view. Since this is not the case the figure is somewhat in error. The general trend is clear however. Inclusion of thc non-failed specimens reveals that the 2-parameter Weibull distribution is not a good descriptor of the data. Also, the dramatic effect of including the non-failed specimens in the parameter estimation is seen. It should be pointed out that the non-failed specimens refer to the low stress levels. Thus, the result may be affected by a threshold stress and an equation like $x_i = (\Delta\sigma_i - \Delta\sigma_{th})^m \cdot N_i$ [3] might provide better results. The threshold stress range for spectrum loading, however, is a spectrum shape dependent random variable [1] and for simplicity is excluded from this example.

In the case of the combined data, there are so many low value non-failed specimens that a proper or even close to proper rank probability analysis of the data is not possible. In order to get some graphical presentation of the data, the "incorrect" form of including only the failed specimen results was used

(Fig. 8). As for the fillet weld specimens, both a LSF on the failed data and an MML estimation on all data was performed. Again, a dramatic difference between the two estimates is seen. The general trend for this data set is that inclusion of the non-failed specimen results in the analysis increases the estimated scatter of the data, i.e., the Weibull exponent decreases.

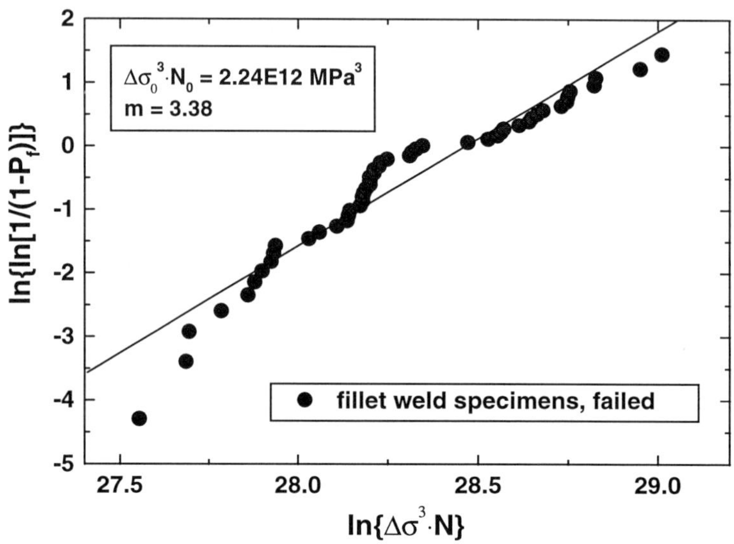

Fig. 6 Rank probability Weibull diagram of failed fillet weld specimens. Solid line refer to LSF of the Weibull plot. m is the Weibull exponent.

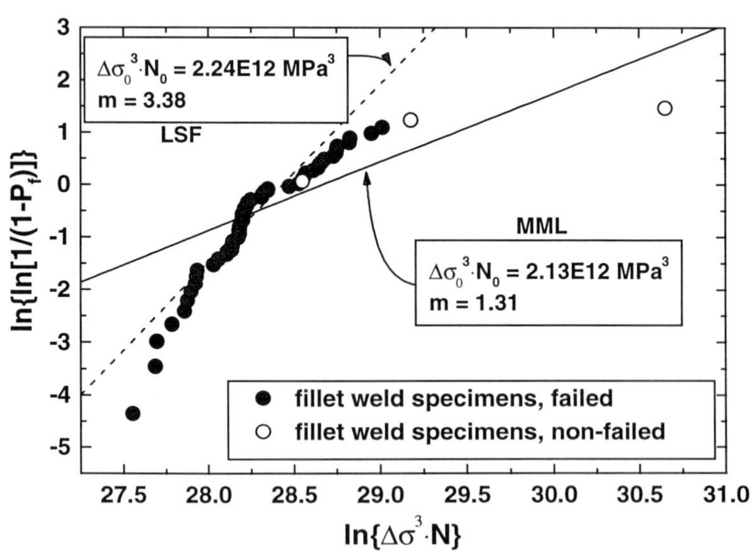

Fig. 7 Rank probability Weibull diagram of all fillet weld specimens. Dashed line refer to LSF of the failed specimens and solid line to MML estimate including all specimens. m is the Weibull exponent.

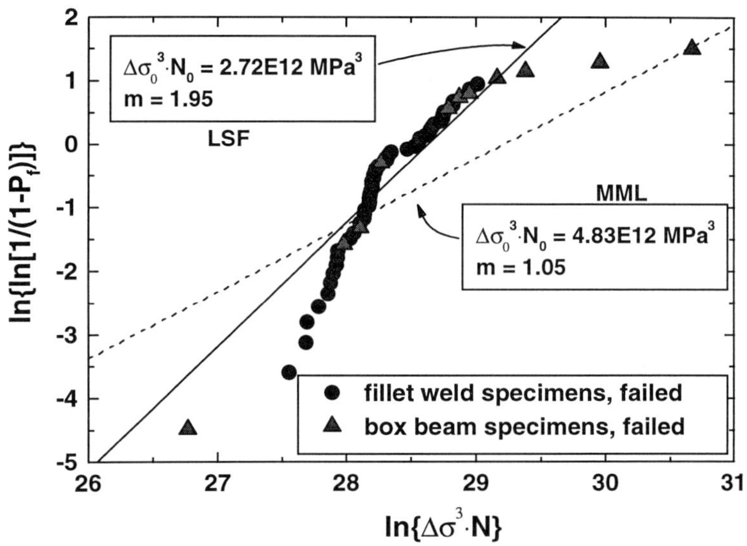

Fig. 8 Rank probability Weibull diagram of all failed specimens. Solid line refer to LSF of the failed specimens and dashed line to MML estimate including all specimens. m is the Weibull exponent.

LOWER BOUND ESTIMATES

The four fits from Figs. 7 & 8 are compared with respect to the estimated mean behaviour in Fig. 9.

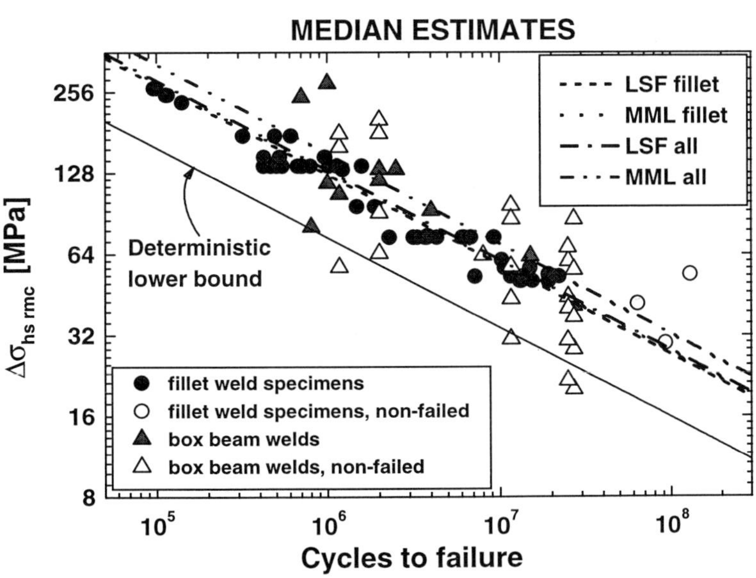

Fig. 9 Comparison of different mean estimates for the fillet weld specimens and the total data set, based on the Weibull distribution. (Compare with Fig. 4).

All LSF fits that are based only on failed samples and the MML estimate of the fillet welds, where the number of non-failed specimens was small produce essentially identical estimates of the mean S-N behaviour (Figs. 3, 4 and 8). The reason for the different behaviour of the MML estimate based on the total data, is due to the large number of high-value non-failures, which both increase the estimate of the mean as well as the scatter.

Similar to the mean estimates, lower bound estimates corresponding to 2.5% failure probability ($-2\cdot\sigma$) are presented in Fig. 10. The Weibull based LSF for the fillet welds (Fig. 10) are seen to produce essentially the same result as the "normal" LSF on all failed specimens (Fig.3). However, the Weibull based LSF gives a more conservative lower bound (nearly the same as the "deterministic" lower bound). The reason for this is that the Weibull distribution lower tail is more conservative than the Lognormal distribution lower tail. This does not however automatically mean that it would be advisable to use the Weibull distribution instead of the Lognormal distribution. Both MML lower bound estimates are similar and clearly more conservative than the LSF estimates. Based on the data, the MML estimates seem actually somewhat over-conservative. The reasons for the possible over-conservatism, may be due to the simplicity of the analysis. It is more than likely, that a fatigue stress threshold should have been included and the use of a Weibull distribution may not be appropriate. A better distribution might be the Lognormal distribution analysed with a randomly censored MML algorithm [5]. A 2-parameter Weibull distribution represents a type of weakest link behaviour. Since S-N data are the sum of three separate events, crack initiation, propagation, and failure, they cannot be considered as representing weakest link behaviour. Even if each separate event would follow a Weibull distribution, the combined distribution would not. This speaks in favour of the Lognormal distribution.

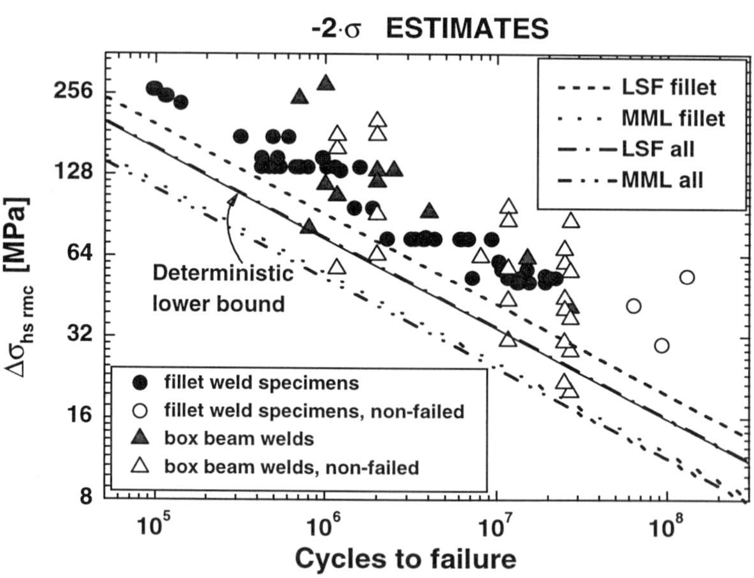

Fig. 10 Comparison of different lower bound estimates for the fillet weld specimens and the total data set, based on the Weibull distribution. (Compare with Fig. 3).

The Weibull distribution was used here as an example because its MML algorithm can be expressed in a closed form (eq. 8). The randomly censored MML algorithm for the Normal or Lognormal distribution is much more complicated and can only be solved numerically.

RANDOMLY CENSORED MAXIMUM LIKELIHOOD ESTIMATION OF THE LOG-NORMAL DISTRIBUTION

There are many indications that a Lognormal distribution describes S-N data better than a Weibull distribution. This is apparently the case also for the present data set (Fig. 11). If the data contain only failed specimens the MML estimate is the same as the LSF estimate for this distribution, but for "randomly censored" data sets the algorithm becomes more complicated [5]. It is not possible to develop a closed form solution. It is, of course, possible to write the expression for the logarithm of the likelihood. It can be expressed as

$$\ln L = -n\ln\sqrt{2\pi\sigma} - \sum_{i=1}^{n}\frac{\delta_i}{2}\left(\frac{\ln\left(\Delta\sigma_i^{\,b}N_i\right)-\ln(m)}{\sigma}\right)^2 + \sum_{i=1}^{n}(1-\delta_i)\ln\left[\int_{\ln\left(\Delta\sigma_i^{\,b}N_i\right)}^{\infty}\exp\left\{-\frac{1}{2}\left(\frac{x-\ln(m)}{\sigma}\right)^2\right\}dx\right] \quad (11)$$

The mean (m) and standard deviation (σ) must solved numerically as to maximise lnL using an appropriate iteration algorithm. One advantage in using equation 11 directly is that it is quite flexible to modifications. The power b (or any alternative form of the relation between Δσ and N) can be estimated simultaneously so this estimate will also be based on the maximum likelihood algorithm. The goodness of different distributions can be compared by comparison of lnL. The distribution producing the highest likelihood is also likely to be the most appropriate for the material.

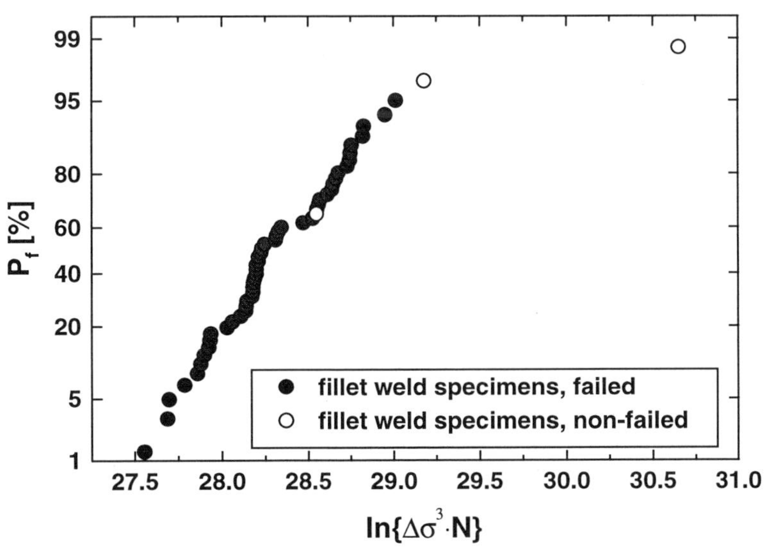

Fig. 11 Rank probability Lognormal distribution diagram of all fillet weld specimens.

SUMMARY AND CONCLUSIONS

In this work, different methods of fitting S-N data have been studied. Special attention has been directed at data sets including non-failed specimens.

Simple least square fitting, involving only the failed specimens, may lead to an underestimation of the true scatter of the data. Non-failed specimens can be included in the analysis by utilizing a maximum likelihood algorithm allowing for random censoring.

The algorithm was derived, as an example, for a 2-parameter Weibull distribution which was used to analyze a specific spectrum loaded S-N data set. The results indicate that the Weibull distribution may not be appropriate for the description of S-N data.

The Lognormal distribution seems to give a better description of the data, but the randomly censored maximum likelihood algorithm for this distribution is more complicated than for the Weibull distribution.

Invoking a threshold stress in the relation between stress amplitude and number of cycles to failure may be beneficial for the analysis.

It is shown that even deterministic lower bound estimates can be attributed specific probability values depending on the sample size. Thus it can be concluded that the probability of success using deterministic reliability is quantifiable.

ACKNOWLEDGEMENTS

This work is a part of the Material Degradation in Reactor Environment project (RAVA) belonging to the Structural Integrity of NPP research programme (RATU2), performed at VTT Manufacturing Technology and financed by the Ministry of Trade and Industry in Finland, the Technical Research Centre of Finland (VTT), the Finnish Centre for Radiation and Nuclear Safety (STUK) and Finnish Nuclear Power industry.

REFERENCES

1 Marquis, G.B. (1995) *High Cycle Spectrum Fatigue of Welded Components.* VTT Publications 240, Technical Research Centre of Finland, Espoo.
2 Wallin, K. (1990). In: *ECF 8 - Fracture Behaviour and Design of Materials and Structures,* D. Firrao (Ed.). EMAS, Warley pp. 1516-1521.
3 Ling, J. and Pan, J. (1997) *Int. J. Fatigue* **19**, 415.
4 Wallin, K. (1989) *Optimized Estimation of the Weibull Distribution Parameters.* VTT Research reports 604, Technical Research Centre of Finland, Espoo.
5 Pascual, F.G. and Meeker, W.Q. (1997) *Journal of Testing and Evaluation, JTEVA* **25**, 292.

FATIGUE LIFE EVALUATION OF GREY CAST IRON MACHINE COMPONENTS UNDER VARIABLE AMPLITUDE LOADING

ROGER RABB
Wärtsilä NSD Corporation, P.O. Box 244, FIN-65101 Vaasa, Finland

ABSTRACT

In a medium speed diesel engine there are some important components, such as the cylinder head, the piston and the cylinder liner, which are subjected to a specific load spectrum consisting of mainly two distinct parts. One is the low cycle part which is due to the temperature field that builds up after that the engine has been started. This low cycle part causes a big stress amplitude but consists of only a couple of thousand cycles during the engine life time. The other part of the load spectrum is the high cycle part due to the firing pressure. The high cycle part has a smaller amplitude but consists of billions of cycles during the engine life time. The cylinder head and the cylinder liner are made of cast iron. In this investigation the true extension into the high cycle domain of the S-N curve for grey cast iron grade 300/ISO 185 was established through fatigue tests with a load spectrum resembling the existing one. This testing resulted in much new and improved knowledge about the fatigue properties of grey cast iron and it was even possible to generalize the outcome of the spectrum fatigue tests into a simple design curve.

KEYWORDS

Fatigue, spectrum load, cumulative damage, S-N curve extension, grey cast iron

INTRODUCTION

The parts of a medium speed diesel engine that are adjacent to the firing chamber will be subjected both to the effect of the high temperature and to the firing pressure. The combined effect results in a specific load spectrum consisting of two distinct parts. The first part, the so called low cycle part, is due to the stress field caused by the temperature field. The temperature field results in a big stress range but the number of starts and stops that determine the number of corresponding stress cycles are limited. In the types of medium speed engines which are treated here, the number of starts and stops during the operational life of the engine is at most about ten to twenty thousands of cycles. The second part of this load spectrum consists of the stress field generated by the firing pressure. The corresponding stress range is lower than the one caused by the temperature field and there are billions of these so called high cycles.

Typical parts of a medium speed diesel engine which are subjected to this kind of load spectrum are the cylinder head, the cylinder liner and the piston. A section of the Wärtsilä Vasa 32LN engine showing these parts is presented in Fig. 1. The cylinder liner is made of centrifugally cast grey cast iron.

The cylinder head is sand cast either of grey cast iron or of nodular cast iron.

The behaviour of metallic materials under the influence of a load spectrum where the high cycle part consists of billions of load cycles is very badly known and documented in the available literature. For load spectrums consisting of at most some millions of cycles, the linear cumulative damage rule of Palmgren-Miner [1, 2] can be successfully used. The well known SAE method [3] is also well suited for load spectrums with a limited number of cycles.

It has long been known that if the low cycle amplitude is above the fatigue limit at its appropriate mean stress this will tend to destroy also the fatigue limit corresponding to the high cycle amplitudes at their appropriate mean stress and cause a situation with cumulative damage. This cumulative damage will occur even when the amplitude of the high cycle load is below the fatigue limit.For load spectrums where the number of load cycles does not exceed the number of cycles needed to reach the fatigue limit by much some simple modifications of the basic S-N curve have been tried as e.g. the method according to H.T. Corten and T.J. Dolan [4]. Erwin Haibach [5] has also suggested a modification of the basic S-N curve where the extension of the S-N curve continues with about double the slope exponent in the high cycle area.

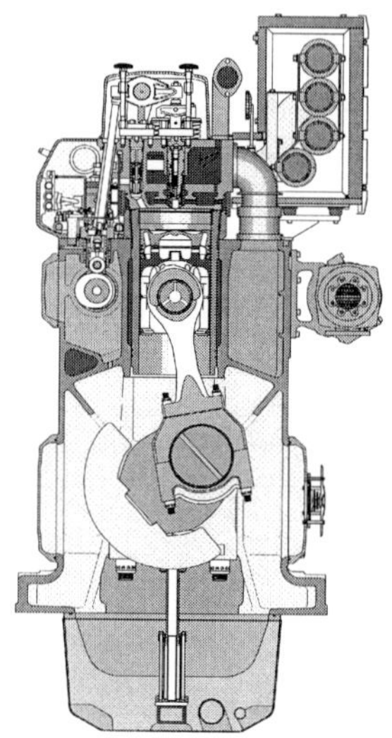

Figure 1. The Wärtsilä Vasa 32LN medium speed diesel engine. The cylinder head, cylinder liner and piston are subjected to both stresses caused by the temperature field and to stresses caused by the firing pressure.

Because of both the lack of understanding of the phenomenon of the low cycle initiated cumulative damage and of some failures that could not be explained as normal fatigue failures due to the constant stress amplitude of the firing pressure, it was decided to investigate this phenomenon with spectrum fatigue tests. It was decided to make these tests on grey cast iron grade 300/ISO 185 specimens made from pieces cut from the flame plate are of the cylinder head as shown in Fig. 1.

CHOICE OF LOAD SPECTRUM

The general appearance of the load history of one block of the spectrum fatigue tests is shown in Fig. 2. The load spectrum generated by this load history is acting on some critical points of the cylinder head and also the cylinder liner. A high stress range arises from the start and stop of the engine. Upon this is the stress amplitude due to the firing pressure superponed. The corresponding load spectrum is in principle determined with the so called Rainflow Cycle Counting [3].

It was decided that it would be the most realistic test if the partial damages at the different high cycle amplitude levels would be kept constant rather than only the ratio of high cycles to low cycles. The original attempt was to choose the ratio of high cycles to low cycles in such a way that the low cycle

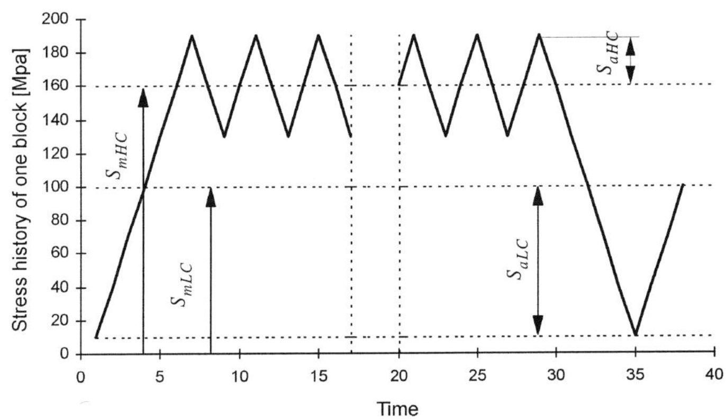

Figure 2. The appearance of one block in the spectrum fatigue test.

damage would be about 25 % and the high cycle damage about 75 % accordingly. The following symbols are used to define the load spectrum:

- S_{mHC} is the mean stress of the high cycle part. This was kept constant and equal to 160 N/mm^2 during the test
- S_{aHC} is the high cycle amplitude. Three different levels below the fatigue limit for the corresponding mean stress were chosen
- n_{HC} is the number of high cycles
- S_{mLC} is the mean stress of the low cycle part of the spectrum
- S_{aLC} is the amplitude of the low cycle part of the spectrum
- n_{LC} is the number of low cycles
- S_0 is a constant prestress of 10 N/mm^2 in the specimens
- N_{cutoff} is the cut off limit chosen to about 10^8 cycles

The following relations connect the high cycle and low cycle parts of the load spectrum:

$$S_{aLC} = \frac{S_{mIIC} + S_{aIIC} - S_0}{2} \quad (1)$$

$$S_{mLC} = \frac{S_{mHC} + S_{aHC} + S_0}{2} \quad (2)$$

Based on what can be found in the available literature it was assumed that the Haigh diagram for the tested grey cast iron grade 300/ ISO 185 would be a straight line between the fatigue limit at fully reversed tension compression to zero amplitude at a mean stress equal to the ultimate tensile strength. In an earlier test on grey cast iron grade 250/ISO 185 it had been found that the fatigue ratio f_R, i.e. the ratio between the fatigue limit in fully reversed tension compression and the ultimate tensile strength is about $f_R = 0.277$. It was assumed that the same fatigue ratio would apply also for grey cast iron grade 300/

ISO 185 which was used in the spectrum fatigue test. Furthermore, it was assumed that the basic S-N curve would reach the fatigue limit S_{fa} at about $N_{fa} = 2 \cdot 10^6$ cycles, and that the extension of the S-N curve into the high cycle domain would have about double the slope exponent according to the suggestion of E. Haibach [5]. Originally the determination of the parameters in the spectrum fatigue test was based on the following data:

$R_m = 339\ \text{N/mm}^2$, the tensile strength as an average from 10 tensile tests

$S_{fa} = f_R R_m = 93.9\ \text{N/mm}^2$, the estimated fatigue limit in fully reversed tension compression

$S_{fa} = 49.6$ -〃-, the estimated fatigue limit with a mean stress equal to $S_m = 160\ \text{N/mm}^2$

The estimated S-N curve and its extension at a mean stress equal to the high cycle mean stress is thus as follows:

$$N_2 = N_{fa}\left(\frac{S_{fa}}{S_a}\right)^k = 2 \cdot 10^6\left(\frac{49.6}{S_a}\right)^{11.305} \text{, the basic S-N curve for } 1 \le N_2 \le 2 \cdot 10^6 \text{ cycles} \tag{3}$$

$$N_{2ext} = N_{fa}\left(\frac{S_{fa}}{S_a}\right)^{2k-1} = 2 \cdot 10^6\left(\frac{49.6}{S_a}\right)^{21.61} \text{, the extension of the S-N curve for } N_2 > 2 \cdot 10^6 \text{ cycles} \tag{4}$$

The estimated S-N curve for a mean stress equal to the low cycle mean stress about 100 N/mm^2 was established in the corresponding way.

$S_{fa} = 66.2\ \text{N/mm}^2$, the estimated fatigue limit with a mean stress equal to $S_m = 100\ \text{N/mm}^2$

$$N_1 = 2 \cdot 10^6\left(\frac{66.2}{S_a}\right)^{11.301} \text{, the basic S-N curve for the low cycle part for } 1 \le N_1 \le 2 \cdot 10^6 \text{ cycles} \tag{5}$$

The required ratio of high cycles to low cycles in the spectrum can be determined with help of the cumulative linear damage rule in the following way:

$$D_1 + D_2 = \frac{n_{LC}}{N_1} + \frac{n_{HC}}{N_{2ext}} = 1.0 \Rightarrow \tag{6}$$

$$\frac{n_{HC}}{n_{LC}} = \frac{N_2 D_2}{N_1 D_1} \tag{7}$$

where $D_1 = 0.25$ is the attempted low cycle damage

$D_2 = 0.75$ is the attempted high cycle damage

The final load spectrum data are summarized in Table 1. The correct values for the high cycle amplitude levels were in fact found only during the test but the ratio of high cycles to low cycles was determined in the way shown above.

Table 1. Spectrum parameters in the spectrum fatigue test of grey cast iron grade 300/ISO 185.

Number of specimens at each level n	High cycle stress in N/mm^2		Low cycle stress in N/mm^2		Cycle ratio in one block
	S_{aHC}	S_{mHC}	S_{aLC}	S_{mLC}	n_{HC}/n_{LC}
5	30.5	160.0	90.2	100.2	5000
5	28.0	160.0	89.0	99.0	27500
5	25.5	160.0	87.8	97.8	200000

THE HAIGH DIAGRAM FOR GREY CAST IRON GRADE 300

Before it was possible to establish the exact stress levels shown in Table 1 for the spectrum fatigue test it was necessary to determine the fatigue limit of grey cast iron grade 300/ISO 185 at least at a mean stress equal to the mean stress 160 N/mm^2 of the high cycle part of the chosen load spectrum. In addition to the obvious need to have an exact knowledge of the fatigue limit, the need to do this fatigue testing was also stressed by an observation made by P.M. Hughes [6] that the test data of grey cast iron appears to define a curved rather than a straight line.

The test specimens were machined from pieces cut from the flame plate area of a cylinder head such as the one on the engine shown in Fig. 1. This cylinder head was sand cast of grey cast iron grade 300/ ISO 185. The appearance of the test specimens are shown in Fig. 3. The same test specimens were also used in the test of the S-N curve and in the actual spectrum fatigue tests.

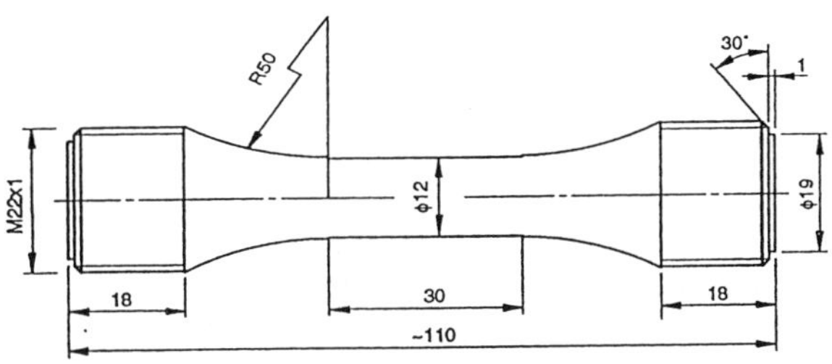

Figure 3. The test specimen used in the fatigue tests on grey cast iron grade 300/ISO 185.

The surface of the specimens was axially polished and the arithmetical mean deviation was about $R_a \leq 1.6$ μm. The ultimate tensile strength was tested on 10 specimens and the average value was R_m = 339 N/mm^2 with a standard deviation of 18.6 N/mm^2. The outcome of the staircase test on these specimens with a mean stress of 160 N/mm^2 is shown in Fig. 4. The cutoff limit in these tests was chosen to 10^7 cycles.

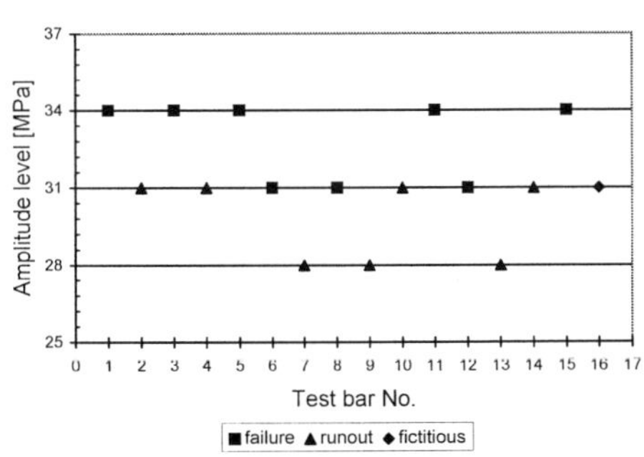

i	f_i	if_i	$i^2 f_i$
2	5	10	20
1	8	8	8
0	3	0	0
Σ	16	18	28
	F	*A*	*B*

Figure 4. Staircase test in tension compression with smooth round specimens of grey cast iron grade 300/ISO 185. Applied mean stress is 160 N/mm^2.

A very good method of evaluating the sample mean and variance from the individual observations in a staircase test is the maximum likelihood estimation according to W.J. Dixon and A.M. Mood [7]. Because of a badly chosen stress increment in this test there are only runouts on the lowest amplitude level and only failures on the highest level. A curve fitting of the most likely density function will therefore deliver a very poor value for the variance in this case. To get an idea of the upper limit of the variance as well, a fictitious observation can be added in the end of the test, as indicated in Fig. 4, and after that the basic equations for a discrete random variable can be applied. According to Dixon and Mood, the most likely density function is found by calculating the maximum value for the following expression:

$$P(n, m|S_a) = K \prod_{i=-\infty}^{\infty} p_i^{n_i} q_i^{m_i} \tag{8}$$

For the tried density function with mean value μ and variance σ^2, the probabilities of failure and runout on the different levels can be calculated with the probability integral.

$$p_i = \frac{1}{\sigma\sqrt{2\pi}} \cdot \int_{-\infty}^{S_{ai}} e^{-\frac{(x-\mu)^2}{2\sigma^2}} dx \quad \text{, the probability of failure on level } i \tag{9}$$

$$q_i = 1 - p_i \quad \text{, the probability of survival on level } i \tag{10}$$

where n_i is the number of failures at the amplitude level S_{ai}
m_i is the number of runouts at this amplitude level
K is a constant that is neither dependent on the sample mean nor on the sample variance

One big advantage of the maximum likelihood evaluation is that there is no need to use a constant stress increment d during the test. Every valid observation will do. By applying these formulas on the test outcome in Fig. 4 the following sample mean and standard deviation are found:

$\bar{S}_a = \mu = 31.1 \text{ N/mm}^2$, sample mean fatigue limit

$s = \sigma = 0.35$ -″- , sample standard deviation of the fatigue limit

The measured value for the fatigue limit is 37.3 % lower than what was expected before this test. The value above for the standard deviation is of course nonsense. With the use of the basic equations for a normally distributed discrete random variable the following upper limit for the standard deviation can be calculated when the fictitious specimen is included:

$$s \le d \cdot \sqrt{\frac{BF - A^2}{(F-1)F}} = 3 \cdot \sqrt{\frac{16 \cdot 28 - 18^2}{(16-1) \cdot 16}} = 2.16 \text{ N/mm}^2 \text{, the sample standard deviation} \tag{11}$$

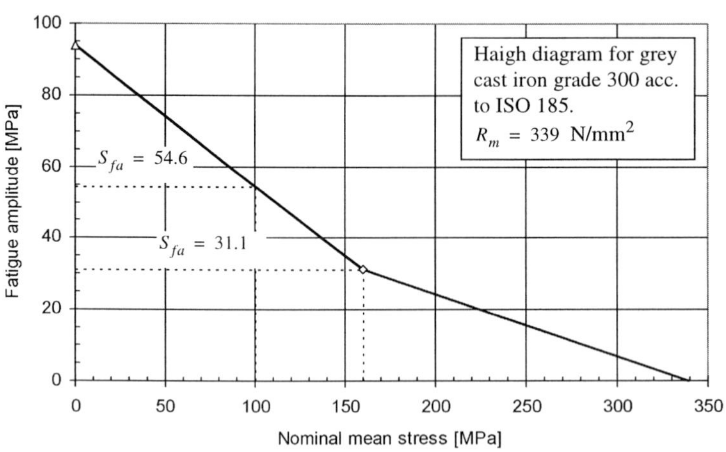

Figure 5. Haigh diagram for grey cast iron grade 300/ISO 185 based on fatigue tests in tension compression of polished round specimens.

To obtain the fatigue limit and standard deviation for the whole population, confidence should be applied to the sample values. For the purpose of establishing the correct amplitude levels in the spectrum fatigue test this question can be left aside. By combining what have been said above it is now possible to sketch the whole Haigh diagram in Fig. 5 for the sample. The two points corresponding to the low cycle and high cycle mean stresses in the spectrum fatigue tests are also indicated.

THE S-N CURVE CORRESPONDING TO THE HIGH CYCLE MEAN STRESS

The S-N curve corresponding to the high cycle mean stress of 160 N/mm^2 had also to be established with fatigue tests in tension compression. Five specimens of the same type as that shown in Fig. 3 were used at three different levels. The outcome of these tests is shown in Fig. 6. In fatigue tests in the finite life area it is usually assumed that the logarithm of life is normally distributed. Because the same method of evaluating the test outcome is also used in the spectrum fatigue tests the basic equations will be given here. The power expression for the desired S-N curve will have the following form:

$$N = N_{fa}\left(\frac{S_{fa}}{S_a}\right)^k \tag{12}$$

If there are n pcs. of observed pairs S_{ai} and N_i then the two unknown parameters N_{fa}, which are the number of cycles at the point where the S-N curve reaches the fatigue limit and the slope exponent k can be solved with the help of the following equation set, which has been derived with the method of least squares:

$$nA - k\sum_{i=1}^{n} lgS_{ai} - \sum_{i=1}^{n} lgN_i = 0 \tag{13}$$

$$A\sum_{i=1}^{n} lgS_{ai} - k\sum_{i=1}^{n} (lgS_{ai})^2 - \sum_{i=1}^{n} lgN_i lgS_{ai} = 0 \tag{14}$$

After solving the equation set (13) and (14) the number of cycles N_{fa} is obtained with the following equation:

$$N_{fa} = 10^{A - klgS_{fa}} \tag{15}$$

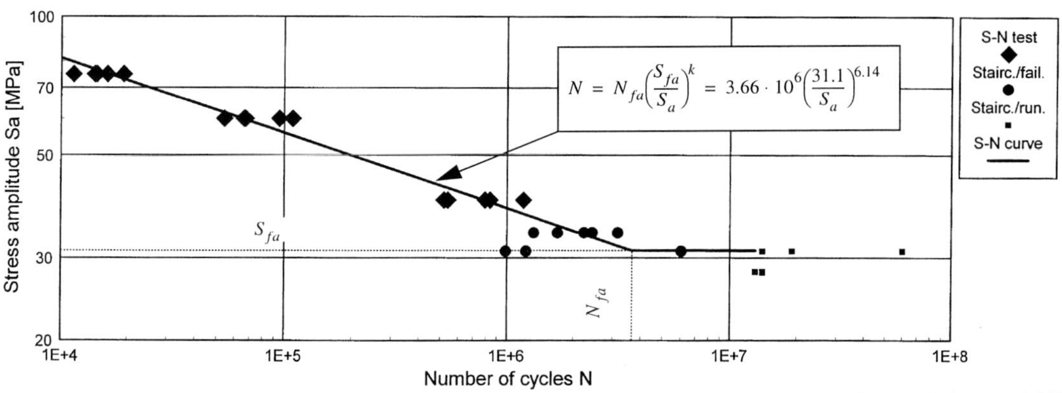

Figure 6. The measured S-N curve for grey cast iron grade 300/ISO 185 at a mean stress of 160 N/mm^2. The observations from the staircase test of the fatigue limit are also included.

As is shown in Fig. 6 the following expression for the S-N curve corresponding to the high cycle mean stress can be derived from the test outcome:

$$N_2 = 3.66 \cdot 10^6 \cdot \left(\frac{31.1}{S_a}\right)^{6.14} \quad , \quad 78.8 \le N_2 \le 3.66 \cdot 10^6 \text{ cycles} \tag{16}$$

The actual S-N curve differs considerably from the original estimate before the test. It was not thought necessary to decide the S-N curve corresponding to the low cycle mean stress at about 100 N/mm^2 with its own fatigue test. This S-N curve was constructed as follows. The fatigue limit at a mean stress of 100 N/mm^2 is $S_{fa} = 54.6$ N/mm^2 according to Fig. 5. It is further assumed that the S-N curve corresponding to the low cycle mean stress starts to slope down from the point where the maximum stress $S_m + S_a$ is equal to the tensile strength R_m at the same cycle number $N_o = 78.8$ cycles as the measured S-N curve corresponding to the high cycle mean stress. Furthermore, it is assumed that both S-N curves reach their fatigue limit at the same cycle number $N_{fa} = 3.66 \cdot 10^6$ cycles. With these assumptions the expression for the S-N curve corresponding to the low cycle mean stress has the following form:

$$N_1 = 3.66 \cdot 10^6 \cdot \left(\frac{54.6}{S_a}\right)^{7.28} \text{, the S-N curve corresponding to a mean stress of about 100 N/mm}^2 \tag{17}$$

This estimate of the S-N curve corresponding to the low cycle mean stress differs also considerably from the original estimate according to eq. (5). To be able to transform the outcome of a fatigue test into reliable design rules it is also important to evaluate the standard deviation of the fatigue life and strength and to apply confidence levels to both the sample mean and the sample standard deviation. For a test of the S-N curve an unbiased estimate of the variance can be calculated from the following expression:

$$s_N^2 = \frac{1}{n-1} \sum_{i=1}^{n} [lgN_i - (A - k \cdot lgS_{ai})]^2 \tag{18}$$

This variance on the logarithm of life can be transformed to a corresponding relative standard deviation for the fatigue strength with the following equation:

$$s_r = 1 - 10^{-s_N/k} \tag{19}$$

In the test of the S-N curve above the following standard deviations could be evaluated from the test data:

$$s_N = 0.124 \Rightarrow s_r = 0.045$$

The sample standard deviation is characterized by the chi-square distribution [8]. By applying a 90 % confidence level to the evaluated sample relative standard deviation a value is obtained for the whole population of which only 10 % of the sample standard deviations would go beyond. For a number of degrees of freedom of $n - 1 = 14$ the unbiased chi-square parameter is $h_1 = 7.790$ and the following relative population standard deviation can be calculated:

$$s_{rC} = \sqrt{\frac{n}{h_1}} \cdot s_r = \sqrt{\frac{15}{7.790}} \cdot 0.045 = 0.062$$

THE SPECTRUM FATIGUE TESTS

After the determination of the fatigue limit and the S-N curve corresponding to the high cycle mean stress the final spectrum test parameters, according to Table 1, could be settled. However, the chosen ratio between the number of high cycles to low cycles in one test block was not changed from the original estimation. A cutoff limit of about 10^8 cycles was used and the same test specimens as shown in Fig. 3 were also used in this spectrum fatigue test on grey cast iron grade 300/ISO 185. The test results are summarized in Table 2.

The outcome of the spectrum fatigue tests was in many respects unexpected and provided much new knowledge. One of the most surprising findings was that the starting point of the extension of the S-N curve suddenly jumps with a factor of about 3 to the left from the end point of the normal S-N curve. Another very important finding was that with increasing number of high cycles, the extended S-N curve tends to become more flat. The outcome of this spectrum fatigue test could be approximated with two distinct branches with the help of the equation set (13) and (14). The evaluated extension of the S-N curve is shown in Fig. 7. A third surprising finding was that a very low number of low cycles will suffice to initiate this cumulative damage. Two of the failed specimens have only 75 low cycles each.

Table 2. The spectrum load fatigue test on grey cast iron grade 300/ ISO 185 with polished round specimens in tension compression.

Ratio of high cycles to low cycles n_{HC}/n_{LC}	High cycle part of the spectrum			Low cycle part of the spectrum		
	Mean stress S_{mHC} N/mm^2	Stress amplitude S_{aHC} N/mm^2	Number of high cycles to failure n_{HC}	Mean stress S_{mLC} N/mm^2	Stress amplitude S_{aLC} N/mm^2	Number of low cycles to failure n_{LC}
5001	159.7	30.45	1.11E+06	100.1	90.1	222
5001	159.7	30.45	3.43E+06	100.1	90.1	685
5001	159.7	30.45	2.20E+06	100.1	90.1	439
5001	160.0	30.50	1.82E+06	100.3	90.3	364
5001	159.5	30.40	9.30E+05	99.9	89.9	186
27501	160.0	28.00	1.08E+07	99.0	89.0	393
27501	160.0	28.00	2.61E+07	99.0	89.0	950
27501	160.0	28.00	3.00E+06	99.0	89.0	109
27501	160.0	28.00	3.71E+06	99.0	89.0	135
27501	160.0	28.00	2.06E+06	99.0	89.0	**75**
200001	160.0	25.50	9.47E+07	97.8	87.8	482
200001	160.0	25.51	1.50E+07	97.8	87.8	**75**
200001	160.0	25.51	1.70E+08 [1]	97.8	87.8	852
200001	160.0	26.50	1.17E+08 [1]	98.2	88.2	586
200001	160.0	26.50	5.30E+07	98.2	88.2	265

[1] runout

The extension of the S-N curve starts from the fatigue limit 31.1 N/mm^2 at $N_{e1} = 1.25 \cdot 10^6$ cycles. The transition to branch 2 takes place at a high cycle amplitude of $S_{e2} = 28.3$ N/mm^2 at $N_{e2} = 4.86 \cdot 10^6$ cycles. The evaluation of the spectrum fatigue tests was made with the equation set (13) and (14) and the following expressions for branch 1 and branch 2 of the extended S-N curve were provided:

$$N_{br1} = N_{e1}\left(\frac{S_{fa}}{S_a}\right)^{k_1} = 1.25 \cdot 10^6 \cdot \left(\frac{31.1}{S_a}\right)^{14.55} \quad , \quad 1.25 \cdot 10^6 \le N_{br1} \le 4.86 \cdot 10^6 \tag{20}$$

with the sample standard deviations $s_N = 0.340$ and $s_r = 0.052$

$$N_{br2} = N_{e2}\left(\frac{S_{e2}}{S_a}\right)^{k_2} = 4.86 \cdot 10^6 \cdot \left(\frac{28.3}{S_a}\right)^{27.78} \quad , \quad N_{br2} > 4.86 \cdot 10^6 \tag{21}$$

with the sample standard deviations $s_N = 0.460$ and $s_r = 0.037$

The true extension of the S-N curve into the high cycle domain differs very much from what was believed before these tests as can be seen by comparing equations (20) and (21) to the estimated one in equation (4). The amount of low cycle damage was surprisingly low as can be seen in Table 3. It was far below the attempted damage $D_1 = 25\ \%$. This finding made it easier to generalize the outcome of the spectrum fatigue tests into a design curve.

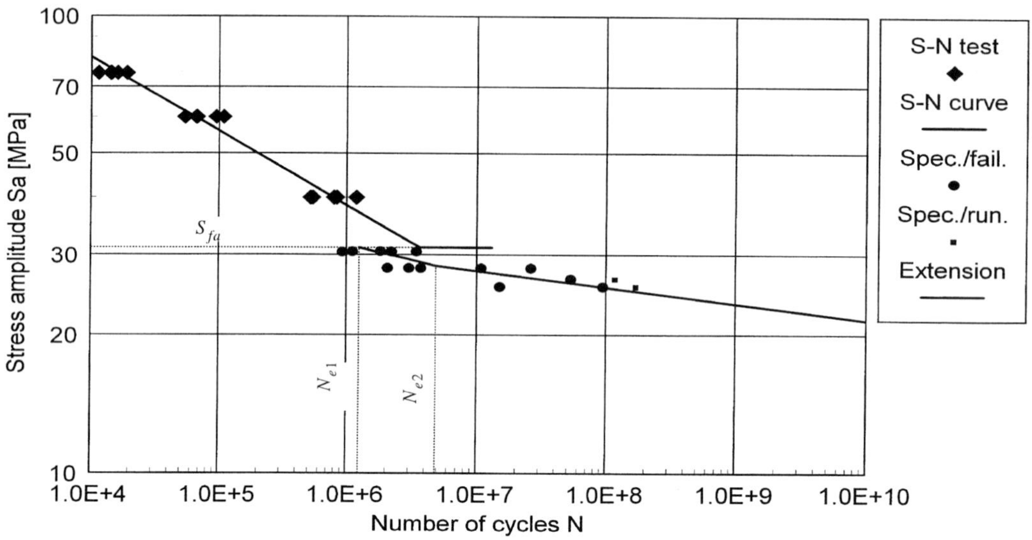

Figure 7. The measured extension of the high cycle S-N curve for grey cast iron grade 300/ISO 185 in a spectrum fatigue test with a high cycle mean stress of 160 N/mm^2. The basic S-N curve is also indicated.

Table 3. The amount of low cycle damage on the three different test levels in the spectrum fatigue tests.

Test level No.	Low cycle amplitude S_{aLC} [N/mm^2]	Average number of low cycles n_{LC} from Table 2	Low cycle life N_1 in cycles acc. to eq. (17)	Low cycle damage $D_1 = n_{LC}/N_1$ [%]
1	90.1	379	95470	0.40
2	89.0	332	104400	0.32
3	87.8	512	115240	0.44

GENERALIZATION OF THE RESULTS INTO A DESIGN CURVE

The results of the spectrum fatigue test can be generalized into a design curve that is easy for the designer to use. This design curve will give the relation between the number of high cycles in the load spectrum and the needed safety factor against the fatigue limit for the high cycle stress amplitude. The starting point is the needed safety factor against the fatigue limit for constant amplitude loadings. The magnitude of the needed safety factor is dependent on the allowed probability of failure and the magnitude of the population standard deviation. The estimation of the population standard deviation is in this case based on the found sample standard deviation for branch 1. The unbiased chi-square parameter for a number of degrees of freedom equal to 9 is with a 90 % confidence $h_1 = 4.168$. The corresponding relative population standard deviation is then as follows:

$$s_{rC} = \sqrt{\frac{n}{h_1}} \cdot s_r = \sqrt{\frac{10}{4.168}} \cdot 0.052 = 0.081 \tag{22}$$

For some other considerations a relative standard deviation for the population of $s_r = 0.09$ has been chosen and this will therefore be a conservative choice. To reduce the mean fatigue limit to the desired probability of failure P it has to be reduced with a certain number λ of standard deviations. This number can be calculated iteratively with the probability integral as follows:

$$P(x \leq \lambda) = \frac{1}{\sqrt{2\pi}} \cdot \int_{-\infty}^{\lambda} e^{-\frac{x^2}{2}} dx \tag{23}$$

The required safety factor can now be calculated with the following equation:

$$S_F = \frac{1}{1 + \lambda s_{rC}} \tag{24}$$

For an allowed failure probability of $P = 10^{-3}$ the corresponding parameter value is $\lambda = -3.09$, which for the above mentioned population standard deviation corresponds to a safety factor of $S_F = 1.39$. If the allowed probability of failure is $P = 10^{-4}$ then $\lambda = -3.72$ and $S_F = 1.50$.

For situations involving low cycle initiated cumulative damage it is possible to derive the following expressions for the needed safety factor as a function of the number of high cycles by use of the equations (20) and (21) for branch 1 and branch 2 of the extended S-N curve:

$$S_F(N) = \left(\frac{N}{N_{e1}}\right)^{1/k_1} \cdot S_F \quad , \quad N_{e1} \leq N \leq N_{e2} \tag{25}$$

$$S_F(N) = \left(\frac{N_{e2}}{N_{e1}}\right)^{1/k_1} \cdot \left(\frac{N}{N_{e2}}\right)^{1/k_2} \cdot S_F \quad , \quad N > N_{e2} \tag{26}$$

These equations for the safety factor are shown in graphic form in Fig. 8 for two often used failure probabilities, $P = 10^{-3}$ and $P = 10^{-4}$.

The question arises whether it is possible to leave the low cycle damage completely out of consideration when the required level of the safety factor is judged also for situations where the amount of low cycle damage is considerably higher than in this case. It can be shown that the influence of the low cycle damage on the required safety factor is very little and that it can usually be neglected for the kind of machine parts and load spectra that have been handled here.

From the Palmgren-Miner cumulative damage rule, equation (6), the following expression for the maximum permissible high cycle amplitude can be derived when also the low cycle damage is taken into account:

$$S_a \leq S_{e2}\left(\frac{N_{e2}}{n_{HC}}\right)^{1/k_2} \cdot (1 - D_1)^{1/k_2} \tag{27}$$

It is easily realized that the factor $(1.0 - D_1)^{1/k_2}$ represents the real decrease of the safety factor, according to the equations (25) and (26), due to the amount of low cycle damage. At the end of the engine life time, the amount of low cycle damage will be about 20 % ... 30 % as can be judged from Table 3. Even a low cycle damage of 30 % will not require any significant reduction of the high cycle amplitude as can be seen by applying equation (27) and it can therefore be neglected.

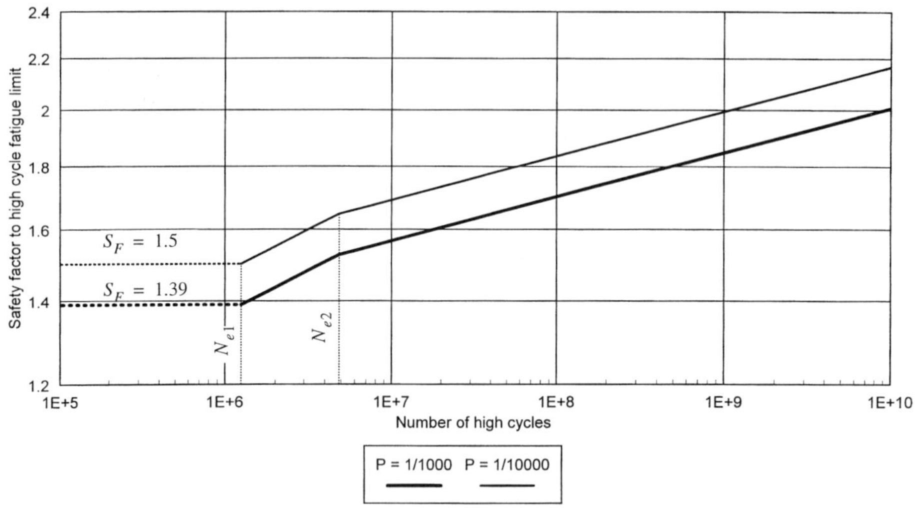

Figure 8. Generalization of the outcome of the spectrum fatigue tests into a design curve with the required safety factor as a function of the number of high cycles and the allowed failure probability.

CONCLUSIONS

The spectrum fatigue tests made on grey cast iron grade 300/ISO 185 resulted in much new and important knowledge. As a side result it was also shown that the Haigh diagram has a positive curvature for positive mean stresses. But the most interesting findings concerned the extension of the S-N curve into the high cycle domain. It was shown that already a few low cycles is enough to initiate the cumulative damage process. The starting point of the extension of the S-N curve jumps to the left with a factor of three and the extension is more flat with an increasing number of high cycles.

However, it was possible to generalize the findings into an easy design curve for the type of load spectra, which were examined. With this curve it is possible to calculate only the high cycle mean stress and amplitude and calculate a safety factor with regard to the fatigue limit corresponding to this mean stress. The level of the required safety factor can be shown to be a function of only the probability of failure and the number of high cycles during the engines life.

It is obvious that it is important to verify the proposed theory in the future with a spectrum test with considerably more low cycle damage because there is the danger that this could increase the slope of the S-N curve extension.

ACKNOWLEDGEMENT

Fatigue tests reported in this paper were performed at VTT Manufacturing Technology. The author would like to thank Jussi Solin who supervised the testing and provided many helpful ideas in preparing this manuscript.

REFERENCES

1. A. Palmgren. Die Lebensdauern von Kugellagern. VDI-Z 69 (1924). Pages 339-341.
2. M. A. Miner. Cumulative damage in fatigue. Trans. ASME. J. Appl. Mech. 12 (1945) A159-A169.
3. Fatigue Design Handbook AE-10. Second edition. Society of Automotive Engineers, Inc. 1988.
4. H. T. Corten and T. J. Dolan. Cumulative fatigue damage. Proc. Int. Conf. Fatigue of Metals, Inst. Mech. London. New York 1956. Pages 235-246.
5. Erwin Haibach. Betriebsfestigkeit. Verfahren und Daten zur Bauteilberechnung. VDI-VERLAG GmbH, Düsseldorf 1989.
6. Document ERC/B.E./11, Issue A Aug 1992. BRITE/EURAM PROJECT 3051-89 ENDDURE. The Application of the Finite Element Method to Fatigue Analysis of Cast Iron Components Containing Stress Concentrations by P.M. Hughes, European Gas Turbines Ltd, Engineering Research Centre.
7. W.J. Dixon and A. M. Mood. A Method for Obtaining and Analyzing Sensitivity Data. Journal of the American Statistical Association 43, pp. 108/126, 1948.
8. J.S. Milton and Jesse C. Arnold. Introduction to Probability and Statistics. Third edition. McGraw-Hill, Inc. 1995.

INCREASE OF RELIABILITY OF ALUMINIUM SPACE-FRAME STRUCTURES BY THE USE OF HYDROFORMED T-FITTINGS

C. Kunz*, M. Schmid**, V. Esslinger**, M.O. Speidel*

* Institute of Metallurgy, Swiss Federal Institute of Technology, ETH Zurich, Switzerland

** Department of Strength&Technology, EMPA Dübendorf, Swiss Federal Laboratories for Material Testing and Research

ABSTRACT

Load supporting structures in transport systems require a particular high standard of safety, reliability and service life time. Combined with the demands for optimised light weight components there is a significant need for new innovative concepts. Aluminium Space-Frame constructions seem to be a very promising way to a) reduce the weight of transport systems and b) to make them more economic [1]. The weak points of this type of construction are always the nodes where the extruded profiles come together. Nowadays, the profiles are most often welded or joined by cast parts in these most critical areas.

It is the aim of this project to replace the cast parts by components made by internal high pressure forming (IHPF) of the same alloy as the profiles. IHPF is the deformation of extruded hollow profiles by simultaneous application of internal high pressure and axial loads.

This publication describes the production of near-net-shape T-fittings by means of IHPF. A comparison of the reliability of conventional welded structures with structures including an IHPF-fitting is given. Mechanical and fatigue properties before and after the IHPF-process are investigated.

The investigations are the result of a cooperation with the companies Alu Menziken Industrie AG, Switzerland and ERNE Fittings GmbH&Co, Austria.

KEYWORDS

Space-frame, hydroforming, internal high pressure forming, T-fitting, welding, light weight construction

INTRODUCTION

Tools for two different shapes are produced. First, an axialsymmetrical geometry with a maximal radial expansion of 100%. By putting additional inserts in the mould it is possible to realise also 10%, 25% and 50% radial expansion (Fig.1a)). By that means it is possible to determine the IHPF forming limits and the change of mechanical and geometrical properties. Different aluminium wrought alloys with different heat treatments are investigated.

In a second step T-fittings are produced. They are welded together with tubes in order to get simple Space-Frame nodes. These parts are tested in respect to reliability and toughness and are compared with conventional welded structures (Fig.1b).

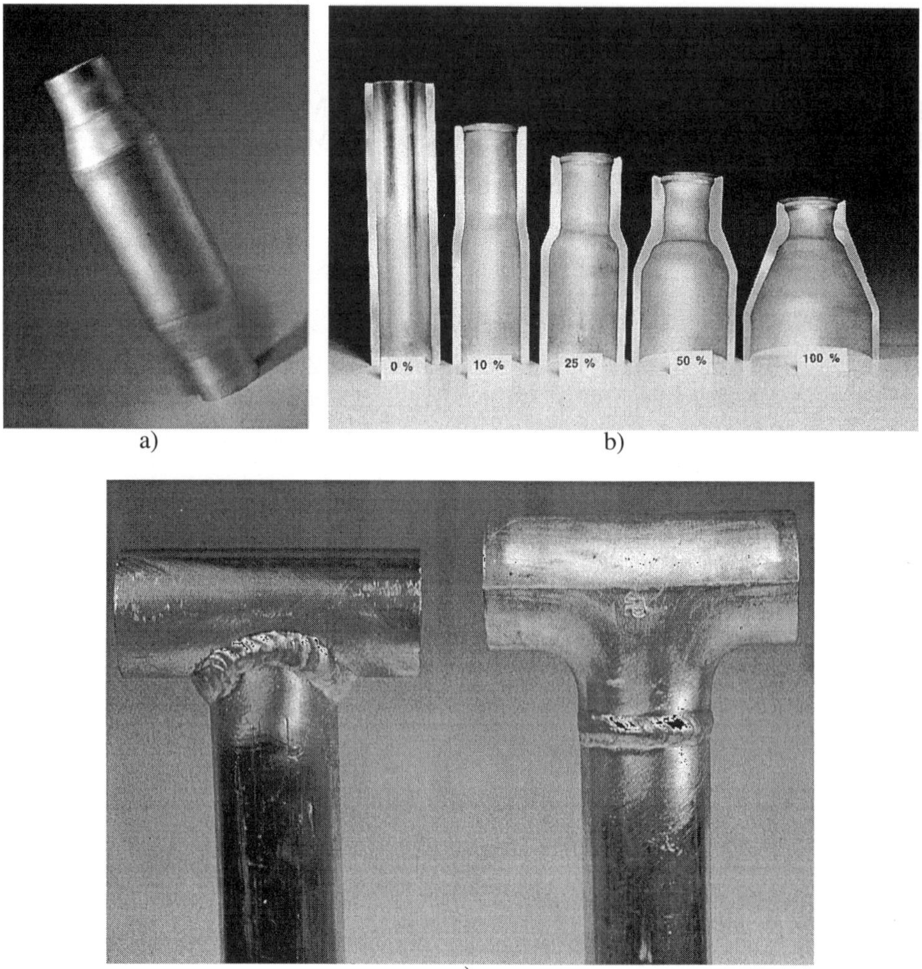

a) b)

c)

Fig.1: a/b) axialsymmetrical expanded tubes of alloy 5005A
c) simple Space-Frame nodes with (right) and without IHPF components (left)

The results of the investigations with the alloys 6060-T1 and 6082-T5 are presented. The diameter of the extruded tubes is 50mm, the thickness of the wall is 6mm.

Other results of this work like internal stresses, mikrostructure, surface quality and ageing behaviour are already published in [2/3].

INTERNAL HIGH PRESSURE FORMING OF T-FITTINGS

With IHPF [4-6] complex near net shape components can be formed using extruded tubes. The deformation of the tubes results of simultaneous acting of internal high pressure and external axial loads. In Fig.2 the IHPF of T-fittings is shown. The surface of the specimen is in contact with the tool during the whole forming process. This and a suitable combination of pressure and loads lead to stress states with a high level of hydrostatic stresses. That extends the forming limits compared to tension dominated stress states. Therefore IHPF allows the formation of more complex hollow parts. Due to the axial compression, material gets into the deformation zones and almost no reduction in the wall thickness occurs. The forming process takes place at room temperature and high degrees of strain hardening lead to excellent mechanical properties. As the production of IHPF components merely takes few process steps, IHPF offers a high level of economy. Inlets can be put into the deformation zones and can be joined during the forming process.

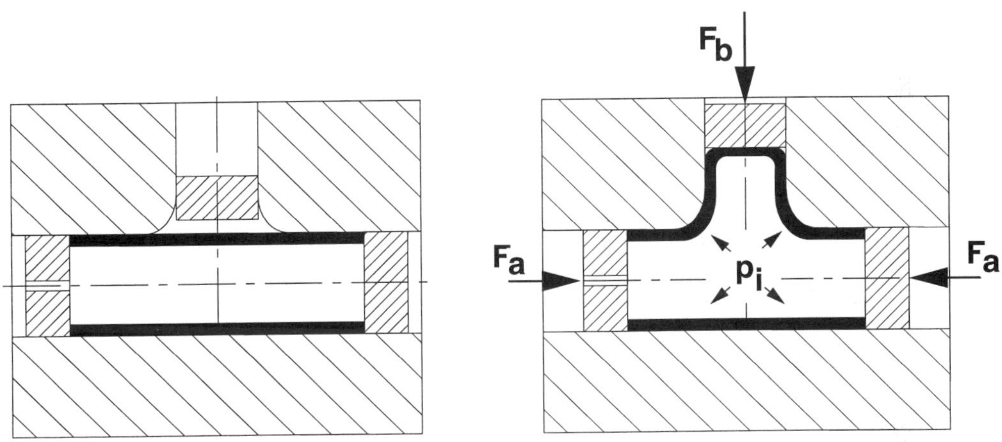

Fig.2: Schematic diagram of the production of a T-fitting with IHPF

AXIALSYMMETRICAL EXPANSION

Considering economical aspects, different thermomechanical heat treatments are investigated. Tubes are expanded in aged tempers (-T1, -T5) and in soft conditions (-W, -O). In the -T1 and -W condition expanded tubes are aged after forming to determine the influence of deformation on ageing behaviour. The mechanical properties are measured before and after IHPF.

Results

Tubes in the -T1 temper are aged at 170°C before and after IHPF. The high dislocation density in the formed components accelerates the ageing process. The time to peak hardness is reduced by one order of magnitude.

The mechanical properties for the different production steps and the maximum gained expansion are shown in table 1. As expected, the IHPF process results in big amounts of work hardening. The strength increases and the toughness decreases. Comparable properties can be found by expanding in the -T1 and the -W conditions.

When expanding annealed tubes (-O condition), additional solution heat treating is required to obtain an acceptable level of strength. In this case unwanted internal quench stresses are brought into the specimen. Ageing 6082 to peak hardness at 170°C after forming in -T1 or -W temper leads not only to higher strength but also to better ductility. The obtained combination corresponds to the properties reached by peak ageing of an undeformed tube.

Table 1. Mechanical properties of different thermomechanical treated specimen

Alloy	thermomechanical treatment *)	Rm, [MPa]	$R_{p0.2}$, [MPa]	A_5, [%]	Ag, [%]	maximal IHPF expansion, [%]
6060	-T1	155	71	26	21	50
	-T1 + IHPF (25%)	225	199	8.5	2.0	
	-W(1h)	118	48	30	24	100
	-W(1h) + IHPF (25%)	173	155	7.7	1.9	
	-O	101	41	29	23	100
	-O + IHPF (25%)	137	121	8	1.6	
	-O + IHPF (25%) + T4	164	82	26	19	
6082	-T1	264	155	23	17	25
	-T1 + IHPF (25%)	308	268	7.9	3.5	
	-T1 + IHPF (25%) + 170°C peak	338	305	11	6	
	-T1 + 170°C peak	338	310	13	8	
	-W(0.1h)	181	67	31	29	
	-W(1h)	201	85	31	29	25
	-W(1h) + IHPF (25%)	311	275	8.6	4.4	
	-W(1h) + IHPF (25%) + 170°C peak	356	333	11	6.5	
	-O	139	54	25	19	50
	-O + IHPF (25%)	158	143	8.6	1.6	

*) IHPF = internal high pressure formed
T4 = solution heat treated + naturally aged

T-FITTINGS

Th goal of the investigation is to compare the fatigue behaviour of conventional welded nodes with components including IHPF-parts.

Experimental Setup

T-fittings are produced according to Fig.2. with maximum possible expansion. The maximal dome height, limited by the mould, is 48mm. The top of the dome is cut of. Dome and end of a 300mm long undeformed tube are prepared for a V-welding and welded together (Fig. 1c)) using a tungsten electrode (TIG-welding).

The conventional construction is built of two tubes that are welded together directly. The welding is in the most critical area regarding stress concentration.

Alloy AlMg5 is used in both constructions as welding addition.

The principle of the used experimental set up is shown in Fig.3. The specimen is fixed at the short ends and tested with cyclic bending loads (R=-1). The fixation of the moving part is realized with a sperical bearing. So no additional bending or torsion stresses are applied during fatigue testing. An electrical motor with an adjustable eccenter sets the apparatus in motion. The applied forces are measured with a load cell. If no failure occurs until 2 million cycles, the test is stopped. The experiments are carried out in air at room temperature.

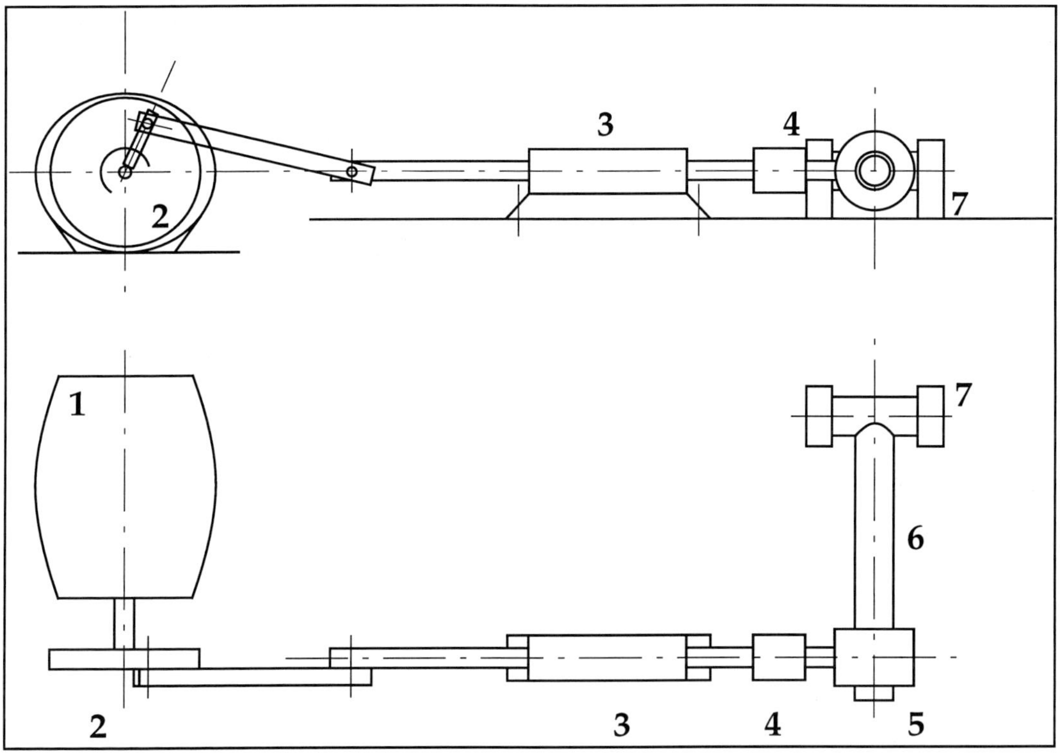

Fig.3: principle of the used experimental setup. 1) motor (12 Hz), 2) eccenter, 3) axial bearing, 4) load cell, 5) spherical bearing, 6) specimen, 7) fixation

The stresses are approximated by calculating a tube loaded with the single force P for bending:

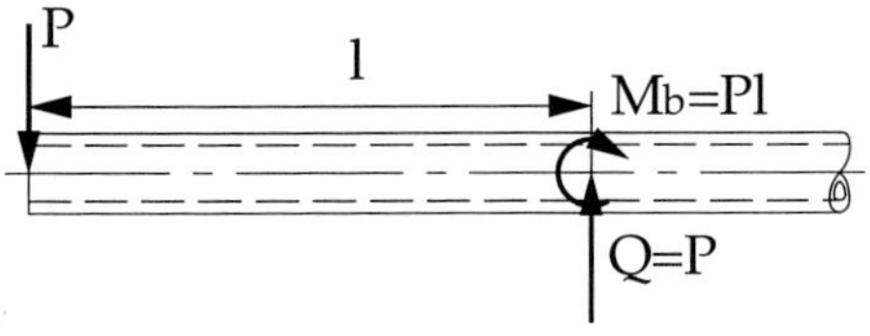

section modulus of a tube: $$W_y = \frac{\pi \left(D^4 - d^4\right)}{32D} \quad (1)$$

maximum stress for bending: $$\sigma_{max} = \frac{M_b}{W_y} \quad (2)$$

The accuracy of this approximation is controlled with strain gauges and was found good. The given stresses are defined in the area of the welding of the conventional constructions. The IHPF-parts are tested with the same forces to get comparable results.

Results

The conventional welded parts fail as expected at the edges of the welding. The cracks in the IHPF-nodes occur in the welding or in the radius of the T-fitting. Only 20% of the IHPF-nodes fail within the welding. In the other 80% the crack initiation is in the radius.

The results of alloy 6060 are shown in Fig.4a) The construction with the IHPF-nodes clearly shows better fatigue behaviour. The stresses to failure are 25-35% higher for the same number of cycles.

For alloy 6082 the difference between the conventional and the IHPF construction is not so evident (Fig.4b)). The reason is the better surface quality of 6060-T1 compared to 6082-T5. But the same tendencies can be observed - the IHPF construction surpasses the conventional construction.

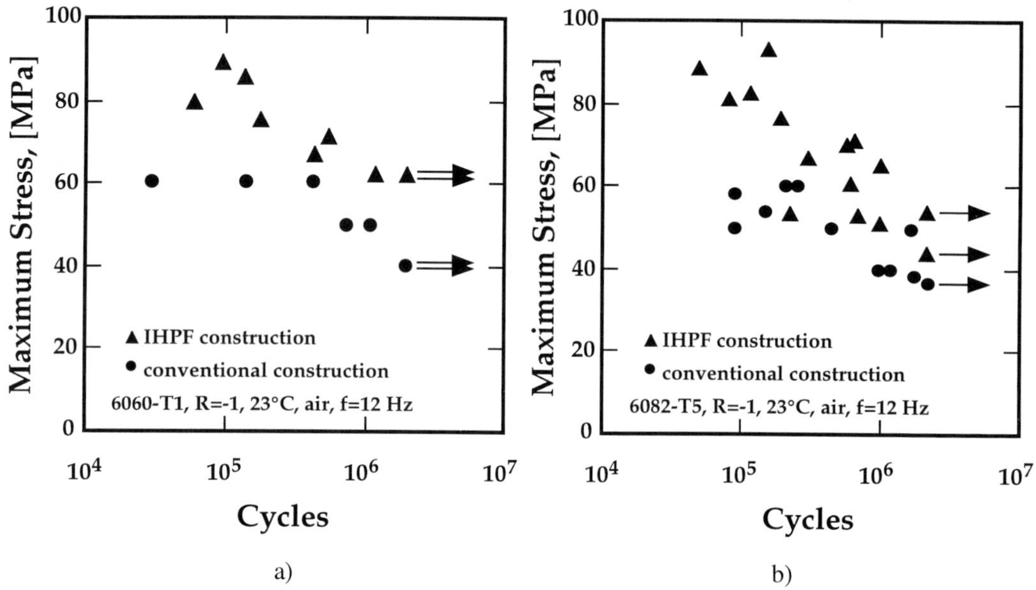

Fig.4: results of the fatigue testing of welded T-nodes (conventional construction vs. IHPF T-fitting) a) 6060-T1, b) 6082-T5

DISCUSSION

IHPF produces stress states with high degrees of hydrostatic components. This leads to high possible degrees of deformation and big amounts of strain hardening. Good combinations of mechanical properties can be obtained by combining IHPF and suitable heat treatments. The best combinations of high degrees of deformation and high strength can be obtained by forming in the freshly quenched condition and subsequent ageing at elevated temperatures. This requires high demands on the chronical sequence of heat treatment and IHPF. The cheapest and easiest way of producing IHPF-parts of 6060 and 6082 is forming them in the naturally aged condition.

The use of IHPF-T-fittings in welded space frame nodes leads to better fatigue behaviour compared to conventional welded constructions. 6060-T1 shows smaller grain sizes near the surface than 6082-T5. This leads to better surface quality after forming and therefore to better fatigue behaviour. The amount of the improvement for 6060-T1 for the IHPF construction compared to the conventional construction is bigger for 6060-T1 than it is for 6082-T5.

In the future it is important to use extruded sections, without coarse grain zones near the surface to avoid bad surface quality what supports the initiation of fatigue cracks. Two investigated constructions will be compared in crash tests regarding energy absorption.

REFERENCES

1. Special Edition ATZ&MTZ, 1995/96.
2. Kunz C., Mächler R., Uggowitzer P.J., Speidel M.O., Conf. Proc. Werkstoffwoche '96, Stuttgart, 1996, 151-156.
3. Kunz C., Mächler R., Uggowitzer P.J., Speidel M.O., DVM-report 122, Berlin, 1996, 223-234.
4. Bitsche E., Renner A., Stapelfeldt G., Blech Rohre Profile, 1/2, 1996, 65-66.
5. Dohmann F., Conf. Proc. Innenhochdruckumformen, Essen, 1997.
6. Mücke K., Blech Rohre Profile, 1, 1995, 17-20.
7. Dohmann F., Bieling P., Blech Bänder Rohre, 5, 1991, 379-385

FATIGUE STRENGTH OF L610-P WING-FUSELAGE ATTACHMENT LUG MADE OF GLARE 2 FIBRE-METAL LAMINATE

A. VAŠEK
Institute of Physics of Materials, Academy of Sciences of the Czech Republic, Brno, Czech Republic

P. DYMÁČEK
Institute of Aerospace Engineering, Technical University of Brno, Czech Republic

L. B. VOGELESANG
Materials and Structures Laboratory, Delft University of Technology, Delft, The Netherlands

ABSTRACT

Flight simulation fatigue test was carried out on a lug specimen representing half of the wing-fuselage attachment lug designed for 60-passenger aircraft L610-P (product of Let Kunovice, Czech Republic). The specimen was made of GLARE 2 fibre-metal laminate with the maximum thickness 18.25 mm in the hole region. Initiation of fatigue cracks was watched and their growth was measured during the test. Crack growth rate was found decreasing with increasing crack length and 100 times lower with respect to crack growth rate in an Al-alloy lug. Fatigue life of the lug subjected to a typical-flight loading history was 10 times longer than a designed life of the aircraft.

KEYWORDS

Flight simulation fatigue test, GLARE 2 fibre-metal laminate, crack growth.

INTRODUCTION

The aircraft L610 being produced by the Czech factory LET Kunovice may be ranged into a new generation of aircraft designed for a short or medium flight range. The L610 can carry up to 40 passengers and has a maximum flight range of 870 km. The maximum take-off weight is 14,500 kg. The aircraft is powered by two General Electric CT7-9 turboprop engines with power 1,320 kW of each. The maximum cruse speed is 490 km per hour. To increase an efficiency of the service of the aircraft the study of the enlarged version of the L610 has been designed [1]. The enlarged version, designated as L610-P, has been designed for 60 passengers with the maximum take-off weight 18,600 kg. Installation of two more powerful turboprop engines Prath & Whittney PW124A with power 1740 kW of each would result in the maximum cruise speed 515 km per hour. The total length of the L610-P is 25.919 m and is by 4.5 m longer then the original version. Wing span has been kept constant. The design of the enlarged version L610-P has been made with the aim to reach the required parameters with minimum changes of the original structure. Besides some other inevitable changes first of all the wing-fuselage attachment structure had to be redesigned because of the longer and more heavy weight-carrying fuselage.

The wing-fuselage attachment structure of the enlarged version L610-P has been designed as an eight-point carrier with four main lug fittings carrying predominantly vertical loads and four smaller supporting lug fittings carrying both vertical and horizontal side loads (Fig. 1). The eight lugs are completed by a shear wall carrying shear loads in the flight direction. As four main lugs are loaded predominantly by tensile loads, the Glare 2 fibre metal laminate with unidirectionally oriented fibres has been applied for their design to reach a damage tolerant and fatigue resistant structural part in the critical joint of the aircraft structure. An improved tensile strength of the Glare attachment fitting lug has been expected as well.

A specimen simulating half of the main attachment fitting lug has been prepared and tested for flight simulation fatigue test.

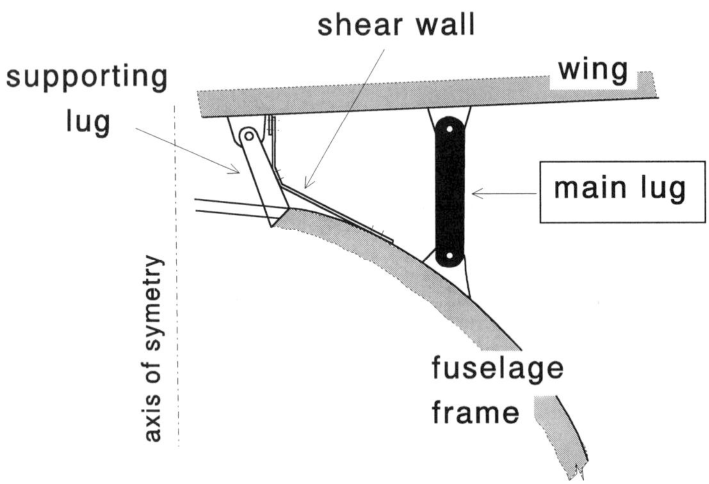

Fig. 1. Location of the main lug in the wing-fuselage attachment structure of the aircraft.

MATERIAL AND DESIGN OF THE LUG

Fibre-Metal Laminates

Fibre-metal laminates are a new type of structural materials, developed at the Delft University of Technology in early eighties [2]. Fibre-metal laminates consist of alternating layers of thin, high strength aluminium alloy sheets (2024-T3 or 7075-T6) and fibre-reinforced adhesive layers (prepregs). The layers are first stacked and then finished in a autoclave hot curing cycle. By now, two types of laminates are commercially available, ARALL and GLARE, with aramid and advanced glass fibres in the prepreg layers, respectively.

Application of the fibre-metal laminates is motivated by their excellent weight savings and mechanical properties (e.g., ultimate strength and fatigue resistance) with respect to the traditional aluminium alloys (Table 1).

Table 1. Mechanical properties of some fibre-metal laminates and 2024 Al-alloy

Property		GLARE 1 (2/1)[1]	GLARE 2 (2/1)	GLARE 3 (2/1)	GLARE 2 (14/13)	Al-alloy 2024-T3
Tensile Ultimate Strength	(MPa)	1077	992	662	1231	455
Tensile Yield Strength	(MPa)	525	347	315	360	360
Tensile Modulus	(GPa)	66	67	60	63	72
Bearing Strength e/D = 1.5	(MPa)	-	566	644	-	758
Bearing Strength e/D = 2.0	(MPa)	834	727	819	-	945
Density	(kg.m^{-3})	2545	2545	2545	2431	2780
Fatigue resistance		good	excellent	good	excellent	poor

[1]Number of Al-alloy/prepreg layers.

Most of the recent research in this material has been focused in flat thin plates. This enables to apply fibre-metal laminates for thin-walled components, like parts of the fuselage or lower wing skin, stabilisers, floor panels and fire walls [3]. References concerning application of multi-layered fibre-metal laminates are rare [4].

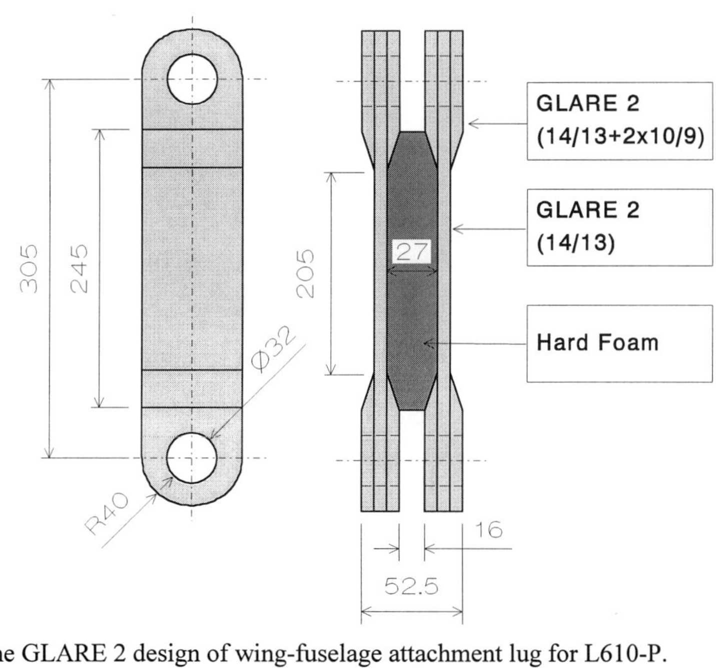

Fig. 2. Sketch of the GLARE 2 design of wing-fuselage attachment lug for L610-P.

Design of the Lug

The lug was designed as to be made of fibre-metal laminate GLARE 2. This type of laminate consists of alternating 0.3 mm thick 2024-T3 Al-alloy sheets and unidirectional glass fibre-epoxy prepreg layers thick 0.25 mm. Two plates of thickness 7.45 mm made of GLARE 2 (14 Al-alloy layers/13 prepreg layers) were intended to carry tensile load applied on the lug. To improve bearing strength of the lug, thickness of the basic plates in the hole area was increased by two additional GLARE 2 plates bonded from both sides on the basic plates. Thickness of the additional plates was 5.25 mm and consists of 10 Al-alloy and 9 prepreg layers. The final thickness of the plate in the hole area was 18.25 mm. To protect buckling instability of the lug structure a space between both basic plates should be filled by a hard foam core. A steel filler of the lug holes is in a real structure highly recommended. Structure of the lug is shown in Fig. 2.

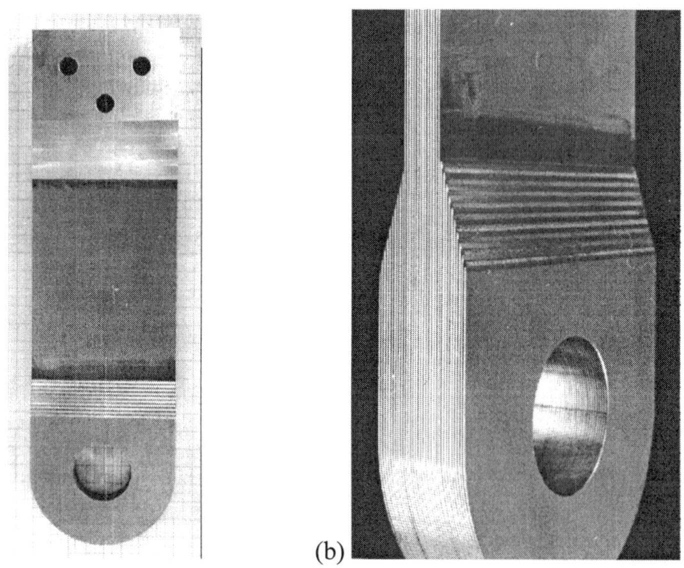

(a) (b)

Fig. 3. GLARE 2 specimen for flight simulation fatigue test. (a) General view of the specimen (length = 280 mm, width = 80 mm). (b) Detail of the lug in the hole region (thickness = 18.25 mm).

For a simplicity the specimen with one lug hole representing only a quarter of the attachment lug was made. The basic plate was stacked with 7 sheets of GLARE 2 laminates each consists of 2 Al-alloy and 1 prepreg layers. The additional plates were stacked with 5 sheets of GLARE 2 (2/1 lay up). The stacked plates were cured in an autoclave at temperature 120°C under pressure 10 bars for 90 minutes. Adhesive film 3M AF-163-2M of thickness 0.15 mm applied for bonding of additional plates on the basic plate was cured in a second autoclave cycle at temperature 115°C under pressure 3 bars for 60 minutes. The semi-product was milled to have the required dimensions. The specimen having the lug hole at one end was provided with two Al-alloy plates bonded on the opposite side of the specimen what enables to fix the specimen into the loading machine properly. The specimen was loaded by a free steel pin inserted in the hole. No filler had been applied in the hole. A fretting damage between the pin and the lug hole was reduced by a lubricant. The specimen is shown in Fig. 3. Total number of Al-alloy and prepreg layers in the thickest part of the specimen (t = 18.25 mm) numbers 34 Al-alloy and 31 prepreg layers.

Required Strength of the Lug

The maximum design tensile load carrying by the main lug was evaluated by means of the FAR 25.561 requirement as F_{max} = 205 kN [1] what is 1.5-multiple of the expected maximum service tensile load ($F_{max,s}$ = 137 kN). Static tensile test of the GLARE lug showed that static ultimate tensile strength was reached at 418 kN what is double of the required maximum design load and three times higher than the maximum expected service load.

Fatigue strength of the lug was required as 100,000 typical flights what is double of the design life of the aircraft. Loading history of the lug in one typical flight of the L610-P version was taken from the full scale test of the L610 version and converted by higher loads carried by the enlarged version. Maximum load in the typical flight carried by one lug was 136.6 kN, minimum load was 80.6 kN and mean load in flight was 105.1 kN. Ground level was taken as 0 kN.

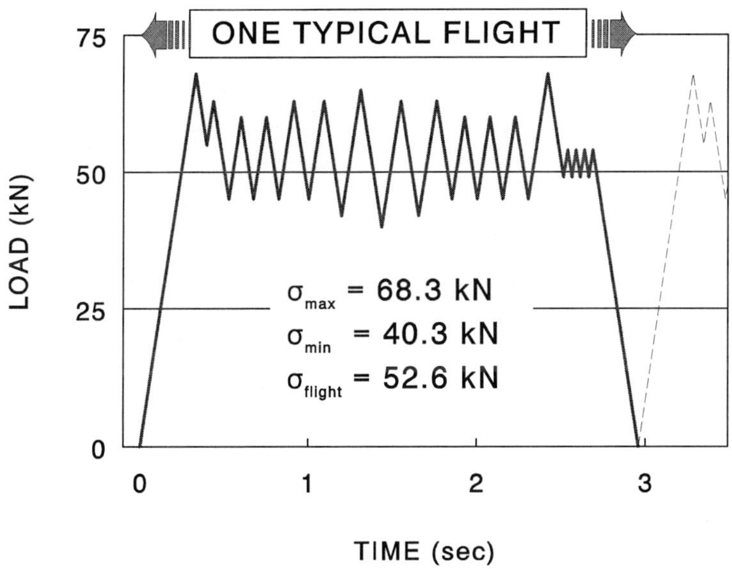

Fig. 4. Part of the loading history applied on the specimen during flight simulation fatigue test as a typical flight.

RESULTS

Fatigue test was performed in a electro-hydraulic testing machine MTS with the maximum load capacity 250 kN controlled by a Fokker Actuator Control System with the maximum load rate 150 $kN.sec^{-1}$. The loading history was simulated as a repeated sequence of 34 peaks and valleys representing one typical flight. As the specimen was designed as a half of the attachment lug maximum load in the loading history was 68.3 kN and mean value was 52.6 kN. A „ground“ valley of 0 kN opened and closed each repetition of the loading block. The simplified and reduced loading history of one typical flight applied on the specimen is drawn in Fig. 4. The loading was periodically interrupted to remove the specimen from the machine and observe the specimen surface in a light microscope. During first 100,000 flights no fatigue crack was found in the specimen. Then more severe loading history with levels of typical flight multiplied by a factor 1.5 was applied for additional 105,000 flights. After 5,000 flights at the sever loading history first fatigue crack was found at the edge of the hole. During loading totally eight fatigue cracks, four in both specimen surfaces, were initiated at the hole edge. The cracks grew radially from the hole edge and at the test termination, after additional 105,000 flights, the longest crack reached length 5.4 mm. Fatigue cracks observed in one specimen surface after test termination are shown in Fig. 5. Surface crack growth rate was maximum at the crack initiation and was evaluated as 9.5 x 10^{-5} $mm.flight^{-1}$. Crack growth rate was decreasing with increasing crack length and at the test termination was equal to 2.9 x 10^{-5} $mm.flight^{-1}$. Growth of surface fatigue cracks versus number of flights is plotted in Fig. 6.

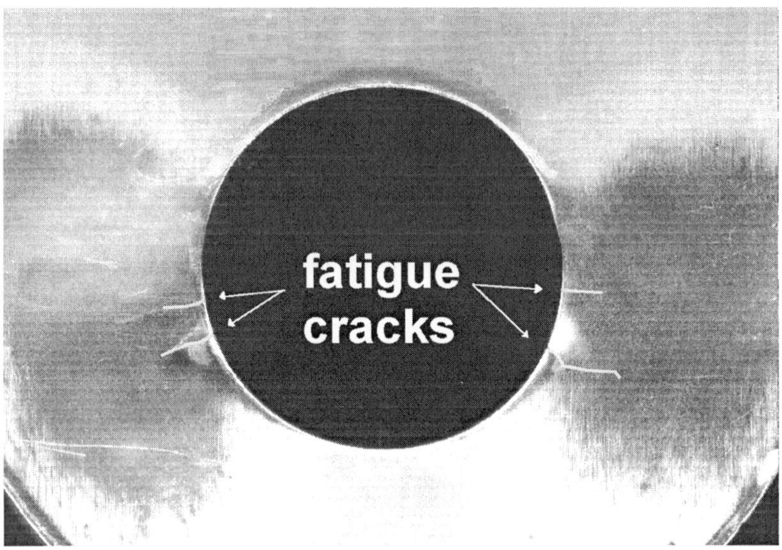

Fig. 5. Fatigue cracks found in the surface at the hole edge of the lug after flight simulation test.

DISCUSSION

Pretty satisfied results were obtained in the flight simulation fatigue test where the required fatigue life, 100,000 typical flights, was easy fulfilled without any cracks found in the material. The lug was able to carry applied fatigue loads of a much severe loading history (levels of typical flights multiplied by a factor 1.5) for additional 105,000 flights after what the maximum crack reached 6 mm length.

The relatively long fatigue life and slow fatigue crack growth in GLARE 2 laminate is due to glass fibres bridging a fatigue crack in the laminate which restrict an opening of the crack and effectively reduce stress intensity factor at the crack tip in metal layers. It results in the extremely low crack growth rate with respect to the rate in a monolithic Al-alloy specimen of a similar configuration. Maximum stress intensity factor of the major crack at the test termination in Al-alloy layers of the GLARE 2 specimen was evaluated by Marrissen's procedure [5] as $K_{max}^{Al\ in\ GLARE} = 8.1\ MPa.m^{1/2}$. Stress intensity factor for a monolithic lug of the identical dimensions and crack length would be three times higher ($K_{max}^{Al} = 24.7\ MPa.m^{1/2}$). If we assume that exponent of v-K curve for 2024-T3 Al alloy is about 4.5, the three times higher stress intensity factor in a monolithic Al-alloy lug would result in a more than hundred times higher crack growth rate with respect to the crack growth rate in the GLARE 2 lug. Moreover crack growth rate in the GLARE 2 lug is decelerating with the increasing crack length. Such a behaviour makes this material extremely resistant against growth of fatigue cracks what offers some important advantages for a service of the lug made from this material. Low and not-accelerating crack growth rate enables to extend significantly inspection intervals of the structure or to eliminate inspections at all.

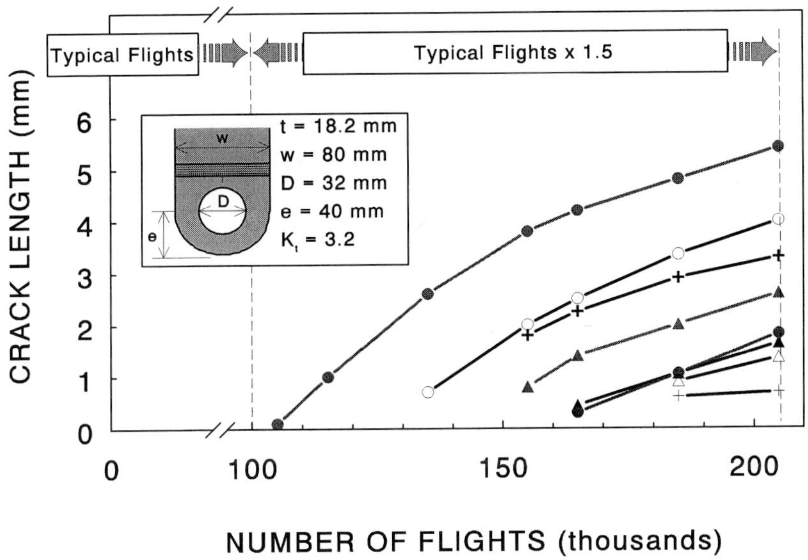

Fig. 6. Growth of fatigue cracks observed in the lug surface during flight simulation fatigue test.

Fatigue life prediction of the lug under the typical flight loading history can be estimated by means of the results obtained earlier at constant amplitude tests of a simple lug made of GLARE 2 [6]. At loading with σ_{max} = 78 MPa which corresponds to the maximum load in the typical flight, the number of cycles in fatigue crack initiation period was 125,000 cycles and 6 mm crack length was reached after 500,000 cycles. Neglecting small vibrations in the flight, the constant amplitude results can be used for an approximate estimit of fatigue life under flight simulation test as almost 500,000 flights.

In order to predict fatigue life in crack initiation period, constant amplitude loading results obtained in open hole specimens has been adopted. It was shown that number of cycles in initiation period of fatigue cracks in notched flat specimens N_i can be estimated using an empirical relation [6]:

$$N_i = (\sigma_n / 5285)^{-4.7} \quad , \tag{1}$$

where σ_n is local stress at the notch root. As GLARE is laminated orthotropic material an evaluation of σ_n had to be done by a finite element calculation procedure. 3D shell elements were used in the system ANSYS to model the orthotropic structure of the laminate. Residual stresses which are present in the laminate layers after hot curing in an autoclave were included in the model as well. Figure 7 shows distribution of equivalent stress in the area around the lug hole at the design load. Maximum stress at the notch root was taken as local stress σ_n. For maximum service loading we received the maximum local stress σ_n = 393 MPa. Neglecting small vibrations during flight, number of flights to initiation can be predicted as number of cycles N_i calculated by equation (1): N_i = 200,000 flights. At the loading with multiplied levels by a factor 1.5, the local stress equals σ_n = 564 MPa and N_i = 36,000 flights. It must be expected that this prediction could be non-conservative due to a more damaging loading of the lug by a pin with respect to loading of open-hole specimens for which equation (1) was derived.

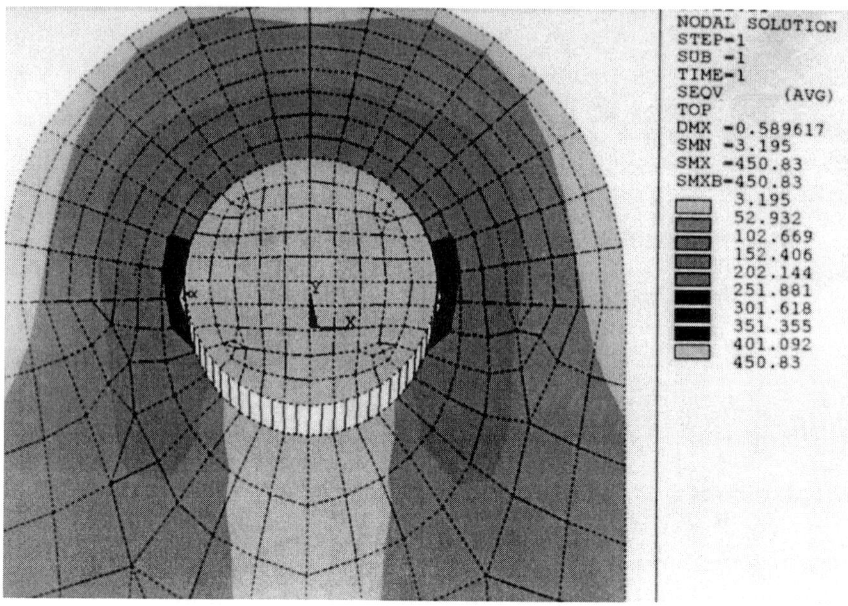

Fig. 7 Result of finite element calculation of equivalent stress in the hole region of the lug loaded by a free steel pin at maximum design load, F_{max} = 102.5 kN. Position of the maximum equivalent stress at the hole notch corresponds to fatigue crack initiation sites.

CONCLUSIONS

Flight simulation fatigue test of the specimen made of GLARE 2 fibre-metal laminate according to a design of the wing-fuselage attachment lug for the Czech aircraft L610-P yields the following results:

- Structure of the lug can carry service loads without any cracks at least for 100,000 repetitions of the typical-flight loading history what is twice longer than the designed life of the aircraft. Application of constant amplitude crack growth data yields an approximate fatigue life prediction of the lug as at least 500,000 flights.
- Application of a more severe loading history (levels of the typical flight were multiplied by a factor 1.5) resulted in a relatively fast fatigue crack initiation, nevertheless crack growth rate was extremely slow and the maximum crack reached 6 mm length after the additional 100.000 more severely simulated flights

ACKNOWLEDGEMENTS

This work was financially supported by the Grant Agency of the Czech Republic (No. 101/96/0319) and by the NATO Scientific Affairs Division (No. HITECH.LG.941331).

REFERENCES

1. Dymáček, P. (1995). Master Thesis. Technical University Brno, Institute of Aerospace Engineering. Brno.
2. Vogelesang, L.B. and Gunning, J.W.: (1986). *Materials and Design*, **7**, 287.
3. Beukers, A., de Jong, T., Sinke, J., Vlot, A. and Vogelesang, L.B. (ed.) (1992). *Fatigue of Aircraft Materials*, Delft University Press, Delft.
4. Mattousch, A. C. (1993). In: *Proceedings of the SAMPE Europe Conference*, Birmingham, 1993.
5. Marissen, R. (1988). *Fatigue Crack Growth in ARALL. A Hybrid Aluminium-Aramid Composite Material*. Report LR-574, Delft University of Technolgy, Delft, the Netherlands.
6. Vašek, A. and Vogelesang, L.B. (1994). In: *Proceedings of the 17th SAMPE Europe Conference*, Basel, 437.

RELIABLE DESIGN OF FATIGUE OF BONDED STEEL SHEET STRUCTURES

H. STENSIÖ, A.MELANDER, A. GUSTAVSSON
Swedish Institute For Metals Research, Drottning Kristinas Väg 48, 114 28 Stockholm, Sweden

G. BJÖRKMAN
Volvo Technological Development, Mechanical Structures, 405 08 Gothenburg, Sweden

ABSTRACT

The interest in adhesively bonded automotive sheet structures is strong due to the potential stiffness and strength increments compared to conventional joining techniques such as for instance resistance spot welding. Naturally, for the conventional joining techniques fatigue design guidelines are well established but are in a development stage for the bonded structures. This paper is a contribution to this process of developing fatigue design procedures for adhesively bonded steel sheet structures.

More specifically, constant amplitude fatigue of peel loaded 1 mm thick sheet steel specimens bonded with an epoxy adhesive is considered. Experimental work as well as two dimensional finite element simulations are presented stressing the importance of adhesive fillet when it comes to stiffness and fatigue strength properties.

KEYWORDS

Fatiguc, Bonding, Fracture Mechanics, Finite Element Method

INTRODUCTION

Structural adhesive bonding is becoming an important joining method in the car industry due to characteristics like high stiffness of the joint and high fatigue performance. The present paper is focusing on design methods and data for adhesive bonding of steel sheet structures. It will be illustrated how methods from fracture mechanics can be used to predict the life of bonded joints and how they can be utilised to analyse effects of the detailed geometry of the joint. It is suggested that based on such procedures practical design methods can be developed for the design of car bodies.

The present paper is devoted to the fatigue performance of adhesive joints which are subjected to peel load. It will specifically be analysed how the degree of filling of the bond influences the fatigue life [2,4]. The paper starts with a presentation of the experimental procedure and results of testing such as stiffness and fatigue life. In the second part a theoretical analyse using fracture mechanics is made, predicting the life of the specimens.

EXPERIMENTAL PROCEDURES

Specimen Configuration

The specimen used in this investigation was a so called T-peel specimen with dimension according to Fig.1.

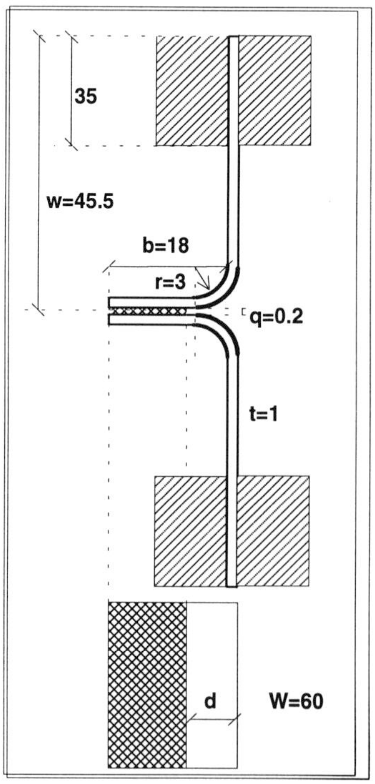

Fig.1. The T-peel Specimen

Typical for this kind of specimen is that the stresses are not distributed uniformly throughout the adhesive but are rather concentrated within a region close to the right edge of the joint (Fig.1). The width of the specimen was 60 mm. The sheet material was a 1 mm thick galvanised rephosphorized low carbon steel with nominal yield strength 220 MPa. The sheets were degreased using heptane and acetone before bonding with a hot-cured toughened epoxy.

The curing took place at 180°C during 37 minutes. Two variants have been tested with different degree of filling of the adhesive, distance d in Fig. 1. They were produced to have fillings of 6 and 8 mm, specimen I and II respectively. Great effort has been taken to manufacture reproducible dimension of the specimen. Therefore Teflon inserts were used both at leading and back edge to achieve a uniform bond line thickness of 0.2 mm and to attain the different degrees of filling. The Teflon spacers remained in place during testing. After testing, the distance d was measured and it was found what specimen I varied between 6 and 7 mm and specimen II between 8 and 9 mm. In the finite element analyses the two larger distances have been used to get the most conservative life estimates.

Testing Conditions and Equipment

Both tensile and fatigue properties of the two specimens were determined. The testing was carried out using a MTS 100kN servo-hydraulic machine available at the Swedish Institute of Metals Research. The tensile tests were run under controlled displacement at a speed of 2 mm/min. The fatigue tests were run sinusoidally at constant-amplitude load levels in tension with a load ratio (R) of 0.05 and a frequency of 20 Hz. Testing was performed in a laboratory environment at room temperature. Load and displacement of the cylinder were registered during both tensile and fatigue testing. Fatigue tests were run until the two parts of the specimen were totally separated, here called total life.

EXPERIMENTAL TEST RESULTS

Tensile Test Results

Table 1 presents the results of the tensile tests. Two specimens of each type were tested and the average maximum tensile load (F_{max}) was evaluated.

Table 1 Maximum loads

Specimen	$F_{max}\,(N)$
I	727
II	603

It was hence found that F_{max} was 17 % higher for specimen I.

Fatigue Test Results

In the fatigue testing, 7 specimen of type I and 8 of type II were tested. During each test the load (P), which is constant, and the cross-head displacement (δ) were registered. The cross-head displacement is a measure of displacement in both machine and specimen. But since the specimen is weak, and thus has a significant deformation, the displacement in the machine is negligible. The stiffness (μ) can consequently be calculated according to

$$\mu = \frac{\Delta P}{\Delta \delta} = \frac{P_{max} - P_{min}}{\delta_{max} - \delta_{min}} \tag{1}$$

A representative graph of the measured stiffness versus load cycles (*N*) is shown in Fig. 2.

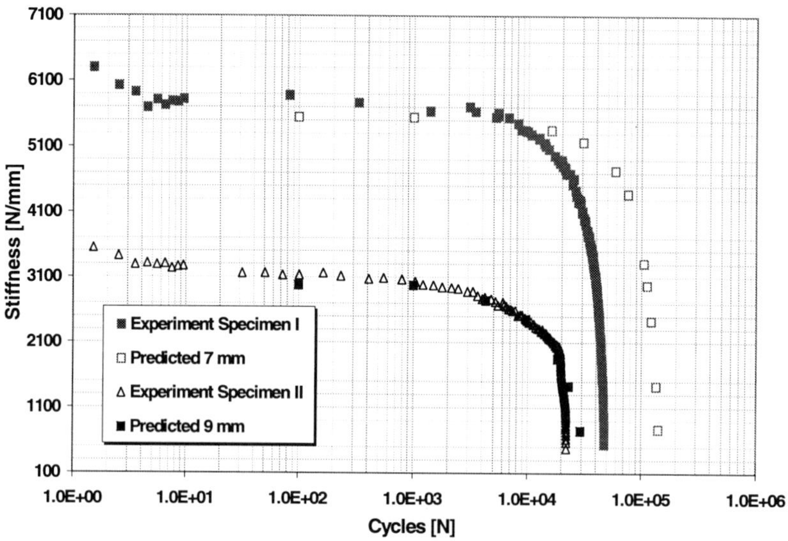

Fig. 2 Stiffness versus cycles from experiment and finite element prediction

As seen in the figure the stiffness takes a constant level during the first 20-30% of the specimen life. This level is here called the initial stiffness. The mean value of these stiffnesses are presented in table 2.

Table 2 Initial Stiffness

Type of specimen	I	II
Initial Stiffness	5500 N/m	3000 N/m

It is worth noting that the initial stiffness is 45% smaller of specimen II than of specimen I that is, when the adhesive starts d = 9 instead of 7 mm from the right hand boundary of the specimen.

The fatigue test results are given as two S-N curves shown in Fig. 3 (solid lines). The fatigue strength is improved with 25% when the adhesive filling is increased 2 mm. Specimen I being the one with longest life.

In bonded joints there are two distinct ways in which a defect can propagate: *adhesive crack growth* occurs along the interface between adhesive and adherent while *cohesive growth* propagates within the adhesive layer. Studying the fracture surfaces of all the specimens it can be seen that the prevailing growth is cohesive.

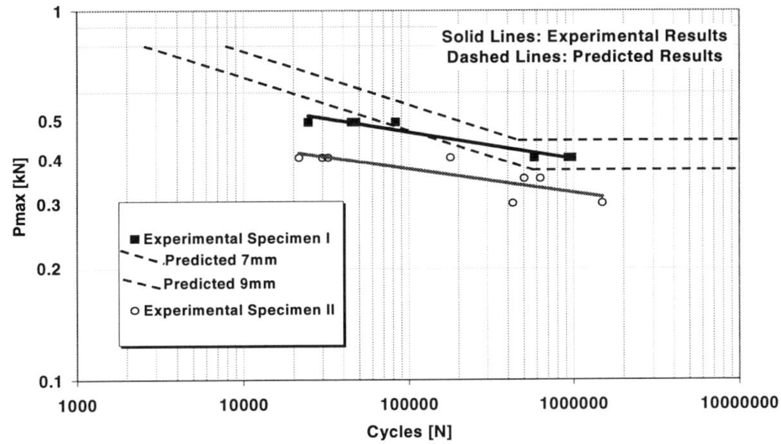

Fig. 3 S-N curves from experiment specimen I and II and predicted from FE analyses.

FINITE ELEMENT ANALYSES

Fracture mechanics is used to predict the fatigue life of the two specimens. In order to do so a Finite Element calculation is performed to predict how the stiffness (μ) and the strain energy release rate (G) varies with crack length (a). Integration of Paris law gives then the life of the specimen expressed in cycles (N).

In figure 4a and b, the finite element model of the structure is shown. The crack propagates through the middle of the adhesive layer, as in the experiments.
The adherents are modelled with 4-node plane stress elements and the adhesive is modelled with 4-node plane strain elements. The Young's modulus and the Poisson's ratio for the adherent material is 210 000 MPa and 0.3 respectively, and for the adhesive 2000 MPa and 0.4 respectively. The code used is ANSYS.

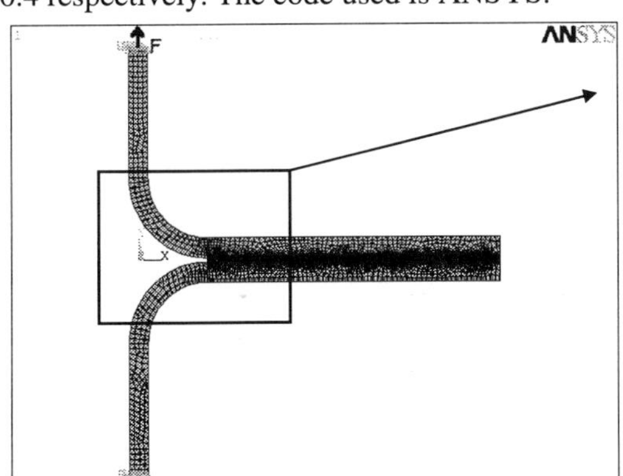

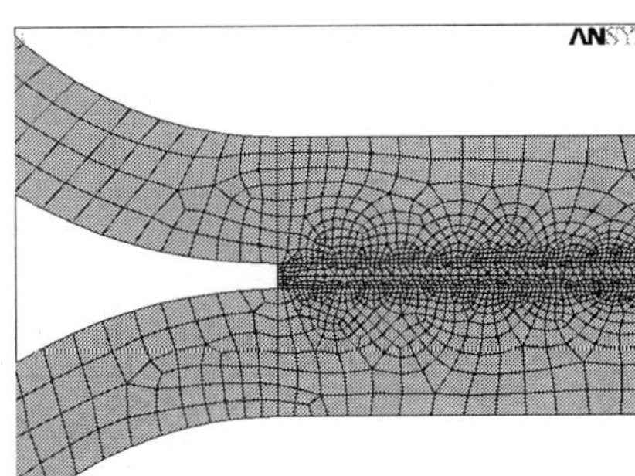

4b

4a

Fig.4 a and b the Finite Element Model

The strain energy release rate (G) and stiffness (μ) are calculated from

$$G = -\frac{dU}{dA} = -\frac{U_{a_{i+1}} - U_{a_i}}{A_{a_{i+1}} - A_{a_i}} \text{ and } \mu = \frac{Force}{displacement} \qquad \text{(2a and b)}$$

where U is the strain energy , A is the crack area at two different crack lengths, a_{i+1} and a_i. Here, $a_{i+1} - a_i = 0.025\,\text{mm}$. This has been done for 16 different crack lengths along the adhesive.

Fig. 5 shows stiffness versus crack length and Fig. 6 strain energy release rate at three different loads versus crack length (a), where the crack length is the distance d in Fig. 1 plus the propagated distance in the adhesive, zero being at the outer side of the sheet. It should from these figures be noticed how sensitive these quantities are to changes in crack length and thus how important the distance d is.

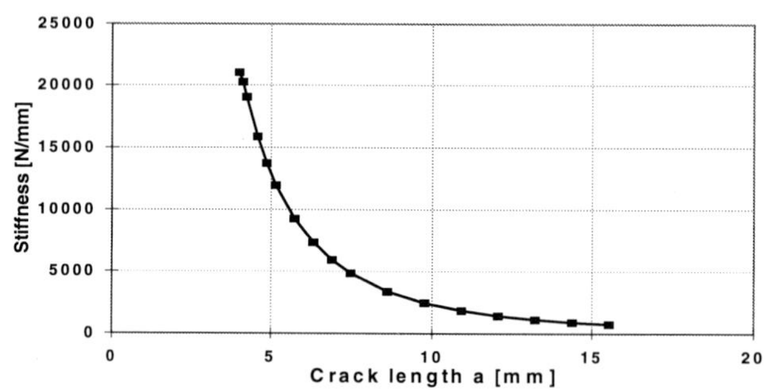

Fig. 5. Stiffness as a function of crack length.

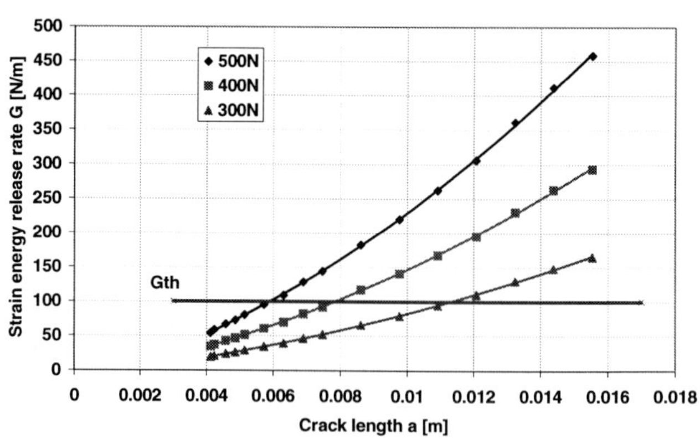

Fig. 6 Energy release rate as a function of crack length for three different loads 500, 400 and 300N.

PREDICTION OF FATIGUE LIFE

In reference [1] the cohesive fatigue crack propagation has been studied for the present epoxy adhesive. The crack propagation rate within this adhesive was found to obey Paris' law.

$$\frac{da}{dN} = C \cdot (\Delta G_I)^n \tag{3}$$

where $\Delta G_I = G_I(P_{max}) - G_I(P_{min})$ refers to the strain energy release rate for mode I with units N/m and da/dN the crack growth rate, m/cycles. $\Delta G_I \approx G_I(P_{max}) = G_{Imax}$ since $R = 0.05$ and thus $G_{Imax} >> G_{Imin}$. The constant C and the exponent n are $5 \cdot 10^{-16}$ and 3.45 respectively. The threshold value was found to be $G_{th} = 100$ N/m according to [1]. Their data do not fully cover the threshold region and therefore the equation above is extrapolated in the region close to G_{th}.

Data for the energy release rate G_{Imax} as a function of crack length a was found in the previous section. When calculating G_{Imax} for different loads it is seen (Fig. 6) that for loads below $P_7 = 440N$ for specimen I and $P_9 = 365N$ for specimen II the energy release rate is less than 100 N/m, that is the threshold value, G_{th}.
It should be noticed from Fig 3. that the loads in the experiments were in-between 500 and 400 N for specimen I (7 mm) and 400 N and 300 N for specimen II (9 mm), i.e. during the testing the loads have been so low that the strain energy release rate was below the threshold value G_{th}.

Integrating $G(a)$ from the initial crack length a_0 of 7 and 9 mm respectively to the final crack length a_f of 17 mm, corresponding to the back edge of the bonded area, gives us the life expressed in cycles (N).

$$N = \int_{a_o}^{a_f} \frac{1}{C} \cdot \frac{1}{-n+1} (G_{Imax})^{-n+1} \cdot \frac{1}{dG_{Imax}/da} da \tag{4}$$

Since loads below P_7 and P_9 give infinite life, the S-N curve will look like the dashed lines in Fig. 3. The solid lines being experimental results.

Hence, using crack propagation data from [1] gives an overestimation of fatigue strength with 30% and infinite life for loads with an energy release rate below the threshold i.e. below P_7 and P_9.

Stiffness as a function of cycles N is also calculated and the result is given in Fig. 2. The predicted initial stiffness, when no or very little crack propagation has taken place, for specimen I is 5500 N/m and for II 3000 N/m, which is equal to the mean values measures in the experiments.

DISCUSSION

As pointed out above it is the crack propagation data near the threshold value that decides the slope of the S-N curve. For the adhesive used in these experiments no such data were found. Only approximate data indicating a threshold value of 100 N/m and crack propagation data between 300 and 1000 N/m. In the calculations above, Paris law was extrapolated dawn to 100 N/m.

In order to improve predictions of fatigue life for these specimens, a simulation has been done by

1. changing crack propagation data in the area governed by Paris law, which we believe could be done within experimental scatter.
2. by allowing a certain propagation below the threshold value. In [3] crack propagation of a structural epoxy was investigated and some crack growth was found below the threshold point of the fatigue curve.

By this fitting procedure the prediction of fatigue life could be greatly improved.

CONCLUSIONS

In this paper a model for predicting fatigue life of an adhesively bonded T-peel specimen has been proposed and compared to experimental results. Fracture mechanics was used and the model was found to qualitatively describe the life of the T-peel specimen. The fatigue strength is given within an accuracy of 30%.

The filling is shown to have a large influence on the fatigue life. Experimental as well as predicted results point out a decrease in initial stiffness of 45% when the bonded area is reduced with one fifth. The corresponding loss in fatigue strength is 20%.

The model of fatigue life shows that the S-N curve is greatly influenced by crack propagation data close to the threshold value. It is therefore of great importance to carefully collect data in this threshold region, to be able to properly predict life of adhesive bonds.

ACKNOWLEDGEMENTS

The present research was financed by the General Research Program of the Swedish Institute for Metals Research and by Volvo Car Corporation which is gratefully acknowledged.

REFERENCES

1. Harries, J.A. and Martin, R.H. (1998), SAE Technical paper no. 980692
2. Gilchrist, M.D. and Smith, R.A. (1993), Int.J.Adhesion and Adhesives, **13**,53-57.
3. Dessureault, M and Spelt, J.K. (1997), Int.J.Adhesion and Adhesives, **17**,183-195.
4. Sheasby, P.G. , Gao, Y. and Wilson, I. (1996), SAE Technical paper no. 960165

ANALYSIS OF STRESS BY THE COMBINATION OF THERMOELASTIC STRESS ANALYZER AND FEM

S.NAGAI
Toyota Motor Co., 1, Toyota-cho, Toyota, Aichi, 471-8572 Japan
T.YOSHIMURA
Toyota Motor Co.
T.NAKAHO
Toyota Motor Co.
Y.MURAKAMI
Department of Mechanical Science and Engineering, Faculty of Engineering, Kyushu University, Hakozaki, Fukuoka, 812-8581 Japan

ABSTRACT

Since the strength of spot welding joints is characterized by the complex stress distribution, it is needed to analyze the stress of the spot welding in detail. However, as an actual construction such as a body of an automobile includes a lot of spot welding joints, it is difficult issue to analyze each of all spot welding joint in the construction. This technical paper describes how to evaluate the stress of spot welding joints both accurately and easily. The method to analyze the stress is based on thermoelastic stress analysis and numerical analysis. Firstly, a first invariable stress J_1 inside Γ, which is an arbitrary domain including the requested spot welding, is measured by the thermoelastic stress analyzer. Secondly, an optimum boundary traction along Γ can be determined by the least square method so that the difference between the measured J_1 and calculated J_1 is minimum. Numerical analysis such as FEM may be used to calculate the stress under the acquired boundary condition. Applied examples using this method that is applicable to analyze the stress distribution of spot welding joints are investigated. One is the specimen's results of spot welding joints under the loading modes of shearing and peeling. The other is the result of an actual automotive construction that have a lot of spot welding joints. As a result, the method is possible to widely apply to the actual automobile construction under the complex stress condition. It is made clear that the appropriate stress distribution can be easily obtained on both the specimen and the actual construction by using this practical method, which has a possibility to reduce both time and cost to ensure the strength of the spot welding such as the body of automobile.

KEYWORDS

spot welding, thermoelastic analysis, finite element method, shearing, peeling, stress invariant, stress component, stress resolution

1. INTRODUCTION

It is one of the most important issue for automotive reliability engineers to ensure the fatigue strength of an actual automobile. Especially enormous time and cost are needed to ensure the fatigue strength of spot welding joints, which are decided by the interior stress condition characterized as singular stress conditions. Furthermore, so far as we concerned, it is the most difficult issue to calculate stress distribution of all spot welding joints on automotive body by using FEM(finite element method), because it needs too large detailed model to calculate complex stress distribution of the spot welding joints. In fact not only calculation using FEM that has detail model but also many kinds of fatigue test are carried out to confirm the fatigue strength of spot welding joints. The method, which was developed by Y.Murakami [(1)], is based on the combination of thermoelastic stress analysis and finite element analysis, and was applied to the simple tensile specimen with one circular hole.

This technical paper investigates an applicable possibility to the spot welding joints that is featured by the complex stress distribution like the actual automotive construction.

2. SUMMARY OF THERMOELASTIC STRESS ANALYSIS

2.1. BASIC PRINCIPLE OF THERMOELASTIC STRESS ANALYSIS

The temperature of a gas increases by adiabatic compression and decreases by adiabatic expansion. Solids behave in a similar manner to gases when subjected to instantaneous compression and tension. This phenomenon is called the thermoelastic effect. Within the elastic deformation of isotropic solids without initial stress, dilatation or contraction(volume change) corresponds to the strain invariant of the first order i_1 ($=\varepsilon_x+\varepsilon_y+\varepsilon_z=\varepsilon_1+\varepsilon_2+\varepsilon_3=\cdots$). Based on Hook s law, i_1 is correlated with the stress invariant of the first order J_1 ($=\sigma_x+\sigma_y+\sigma_z=\sigma_1+\sigma_2+\sigma_3=\cdots$). Thus the heat generation in the elastic range of deformation of an elastic body under adiabatic loading can be proportionally correlated with J_1 and the relationship is given by ;

$$\Delta T = -k T J_1 \qquad (1)$$

where ΔT = temperature change, k = thermoelastic coefficient, T = absolute temperature $^\circ$K, J_1 = the stress invariant of the first order. Thus, J_1 can be determined by the accurate measurement of ΔT.

2.2 METHOD OF STRESS COMPONENT RESOLUTION

In the past, the following two disadvantages for thermoelastic stress analysis have been pointed out.

- Only the stress invariant of the first order J_1 $(=\sigma_x+\sigma_y+\sigma_z=\sigma_1+\sigma_2+\sigma_3=\cdots)$ can be measured and the resolution of all stress components was thought to be impossible in principle.
- Only the information on the surface can be obtained.

However, a method to overcome both these disadvantages has been developed by Y.Murakami. The summary of this method shows as follows.

In order to explain the basic principle of stress component resolution from the measured values of stress invariant, consider a two-dimensional plane stress problem. Fig.1 shows a plate containing a hole. Assume that this plate is fixed along part of its boundary Γ and is subject to external loading along another part of Γ. Thus, we suppose the shape of the plate is known but the load boundary conditions are unknown.

We measure by thermoelastic stress analysis the stress invariant, J_1 of the first order at M points on the surface of this plate. We denote J_1 at point i (i = 1~M) by J_{1i}. If we can determine the load boundary condition from the measured values of J_{1i}, we can determine all stress components at any point inside the plate by numerical stress analysis such as FEM and BEM. The procedure to determine the load boundary conditions follows.

As shown in Fig.2, choose sufficient points along the boundary Γ. A unit normal force and then a unit tangential force is applied to each point in turn (from j = 1 to j = N). There is no definite rule to choose the N points along Γ. However, it is clear that the N points must be chosen so that they fairly reflect the boundary condition.

The next step is to calculate the value of stress invariant J_{1ij} at point i produced by the unit force at point j.

The true boundary condition P_j along Γ should make the following difference zero.

$$e_i = J_{1i} - \sum_{i=1}^{N} J_{1ij} P_j, \quad i = 1\sim M,\ j = 1\sim N. \tag{2}$$

To find the optimum values of P_j in a rational way, we apply the least square method to e_i. We define S as the sum of e_i by

$$S = \sum_{i=1}^{M} (e_i). \tag{3}$$

Thus, the optimum values of P_j must satisfy

$$\frac{\partial S}{\partial P_j} = 0 \tag{4}$$

Equation (4) is a set of N linear simultaneous equation. Solving it, we can determine P_j(j=1~N). Once the values P_j are determined we can calculate the stress components at any point on the plate by superimposing the stress components induced by P_j along Γ. For the calculation of J_{1ij}, FEM or BEM may be used. It has been noted that choosing M and N such that they satisfy the relationship $M>2N$, obtains the best results.

There is no substantial difference between 2D and 3D analysis in this method. In the case of 3D analysis, the interior stress components in 3D body can be determined only from the information of the stress invariant on surface.

3. RESULTS

In order to investigate an applicable possibility for the spot welding joints of the actual automobile, typical spot welding joints under the representative loading modes should be examined by specimens.

3.1 RESULT OF SPECIMEN UNDER SHEARING LOAD

Fig.3 shows a shape of specimen including one spot welding under the shear loading. The material of the specimen is steel (SPCC,JIS G 3141:Japanese Industrial Standard) and the thickness is 1mm. The diameter of the spot welding is about 6mm, and the load range of testing (Δ P) is 1800N. The surface temperature change (Δ T) of an oblique line area (e.g.15×15mm) showed in Fig.3 is measured by the thermoelastic stress analyzer JTG-8010. Table 1 shows the main specifications of JTG-8010. The first invariable stress J_{1I} of measured area is determined by putting measured value Δ T into above equation (1). Fig.4 indicates one of the measurement stress (J_{1i}) on the surface of the specimen. The next step is to calculate the true boundary condition (P_{ij}) as an inverse problem which has already explained at section 2 (cf. equation (3) (4) (5)). As the value P_{ij} is determined, final step is to calculate the stress of specimen by using FEM. The stress analyzing process in this case is carried out by using NASTRAN of FEM software. FEM model is constructed by solid elements in Fig.5. A whole element number is 8554, a node number is 10537, and 795 second CPU time is needed by using Cray computer T94. Fig.6 illustrates the solution of the principal stress value and direction, and Fig.7 shows the interior stress of the spot welding that decides the fatigue life.

Fig.8 shows the comparison between the value of measurement J_1 and calculation on the surface. As the difference value of the stress between measurement and calculation is less than 5%, the stress distribution of the spot welding can be well estimated by using this method.

3.2 RESULT OF SPECIMEN UNDER PEELING LOAD

Fig.9 shows a shape of specimen including two spot welding joints under peeling load. The material of the specimen is steel (SPCC) and the thickness is 1mm. The diameter of the spot weldings are same size as above, but the load range of testing (Δ P) is 200N. As same procedure as above is enforced to analyze the stress, and an example of measurement results is shown in Fig.10. Fig.11 indicates the difference between measurement stress value J_1 and calculation value on the surface in the case of peeling load. An good approximation is also obtained in the case of peeling load.

From the results of the typical spot welding joints specimen under the representative loading modes, it is made obvious that the method is sufficiently efficient to apply to the simple specimens which include only one or two spot welding.

3.3 RESULT OF AUTOMOBILE PARTS

A testing situation is shown in Fig.12, and testing load range (Δ P) is 2000N. The macro stress condition measured by thermoelastic stress analyzer about the transport hook fixture is shown in Fig.13. Fig14 indicates one example of the detail stress condition on the spot welding which is picked up because the stress is the highest in all spot weldings. The result of comparing the calculation result on the surface with the measurement J_1 is shown in Fig.15, where the calculation model is quite equal with the model used in the calculation on specimen.

As a result, this method can give an accurate stress distribution of the spot welding on the actual automobile part under the combination loads. The difference between calculation and measurement value (about 15%) is larger than the result obtained by the specimen as mentioned section 3. The reason is assumed to be caused by the error of revising the large movement in this test and so on. Those problems will be solved to improve the measurement technique through accumulating measurement know-how in the future.

4. DISCUSSION

In the case of estimations of the fatigue strength of spot welding there are several difficult issues as follows. Firstly, as the actual body of automobile has thousands of spot welding joints, it is too hard to construct the detail FEM model of all the body including spot welding joints on a development phase and also to measure the surface stress of every spot welding. Secondly, as the actual complex shaped body is considered to be under the combination loads of shearing and peering, there is no appropriate S-N curve to estimate the fatigue life from the stress on the surface. One means to solve this problem is to arrange the fatigue strength by the interior stress of the spot welding.[(2)] But it is too difficult to measure the interior stress of the spot welding.

A summarized procedure of our method is shown in Fig.16. Our method that has an possibility to resolve those problems needs only standardized detail model (e.g. Fig.5). This model would be used without modification to every spot welding which is requested to analyze. As a result, not only the principal stress value and direction but also interior stress of the spot welding can be obtained both accurately and easily. Furthermore from the database, that is based on the interior stress, we will be able to accurately assume the fatigue life of the spot welding.

Although this method, that is compared with durability test, has a possibility to reduce the cost and time of automobile development, some problems are still remained. Firstly, it is necessary to develop the analyzing method under multi axial loading that have phase difference. This is an unavoidable issue to substitute an actual automobile durability test for our method. A more high technique for revising location on using the thermoelastic analyzer and a reasonable constraint condition on using FEM should be required. Secondly, it is necessary to develop the criteria that is used to predict the fatigue crack initiation. We would like to make efforts to resolve those problems continuously.

5. CONCLUSION

(1) The analyzing method of combining the thermoelastic stress analysis and FEM is possible to apply to the construction under the complex stress condition.
(2) Stress distribution of the spot welding can be obtained by using above method both accurately and easily.
(3) This method has a possibility to reduce the cost and time to evaluate the fatigue strength of spot welding of the automobile.

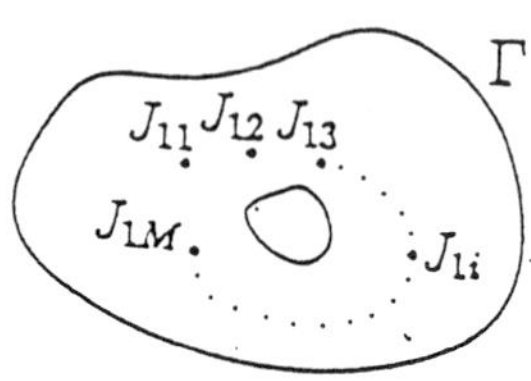

Fig.1 Measurement of J_{1i}

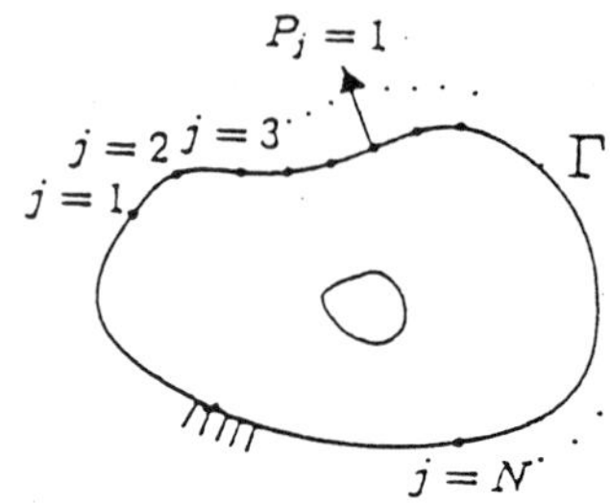

Fig.2 Application of unit traction along boundary

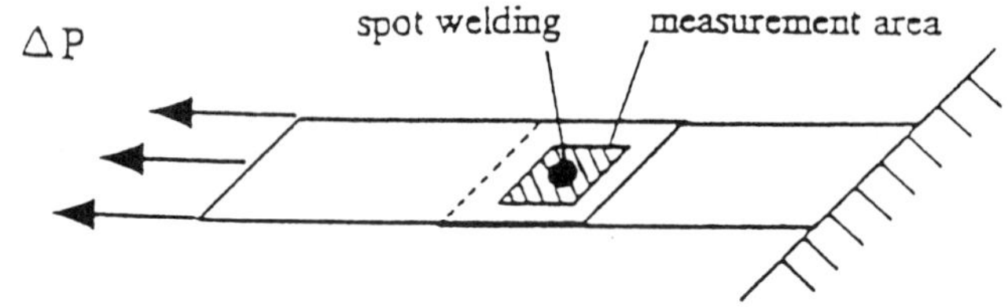

Fig.3 Experimental and analytical condition

Table 1. Main specification of JTG-8010

Stress measurement range	-2000～2000N/mm^2 (-5℃～5℃)
Stress resolvability	0.4N/mm^2 (0.001℃:In case of aluminum)
Measurable frequency	0.5～90Hz
Angle of view	Horizontal 30° Vertical 28°
Horizontal resolution	Over 300 lines (1.7mrad)
Number of display dot	Horizontal 512×Vertical 480
Data memories	512×480×32bit 2pages
Focal range	20 cm～∞ in front of camera

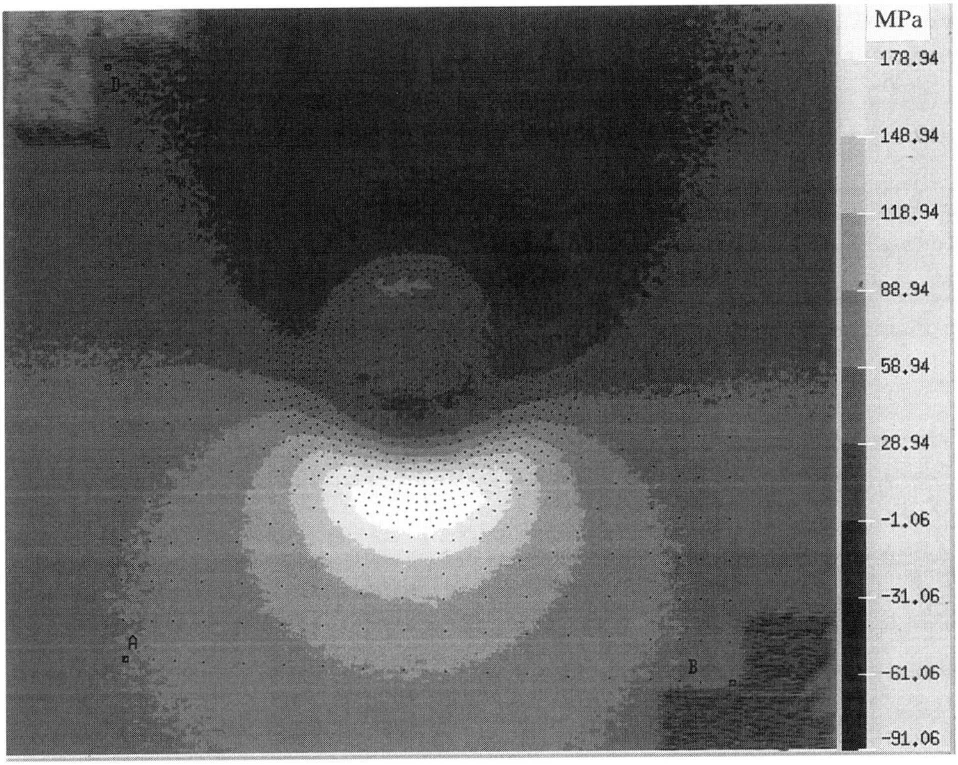

Fig.4 Thermoelastic image analysis on the part of the surface of Fig.3

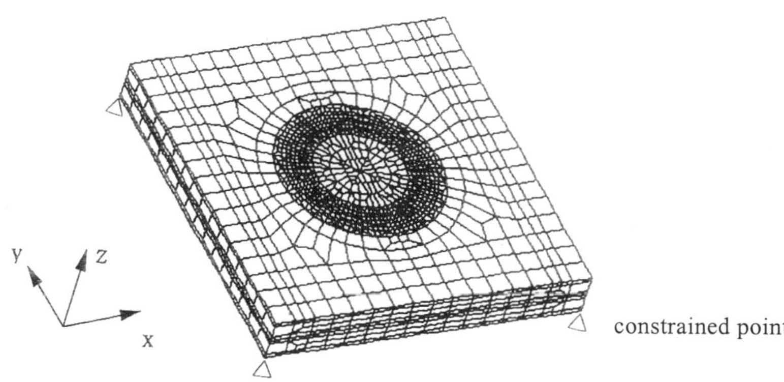

Fig.5 FEM mesh of 3-dimentional model

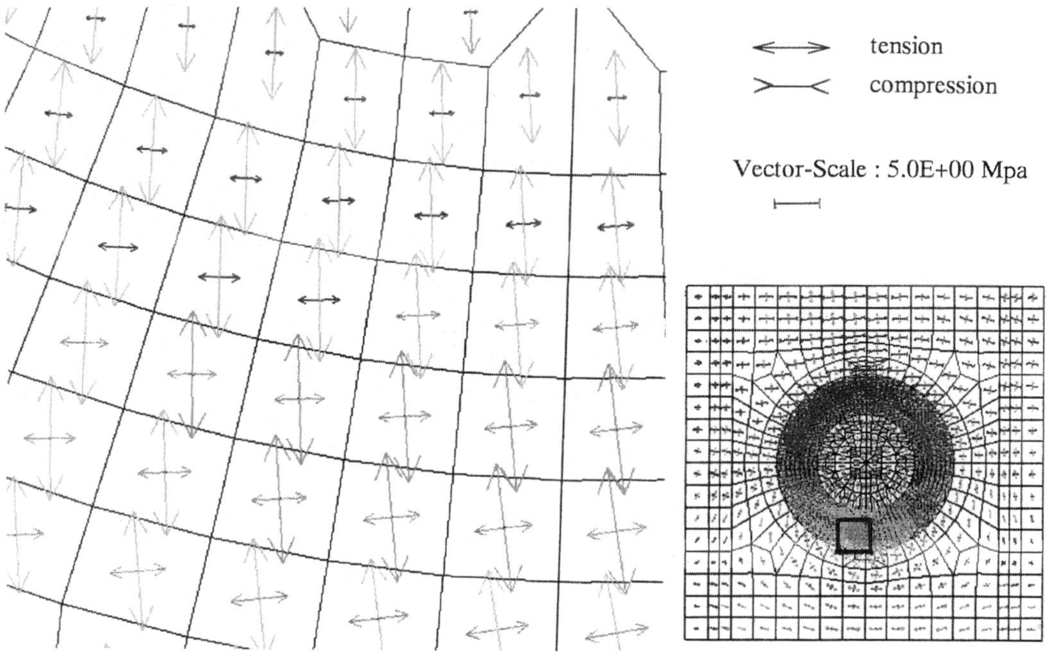

Fig.6 Stress condition on the surface (FEM)

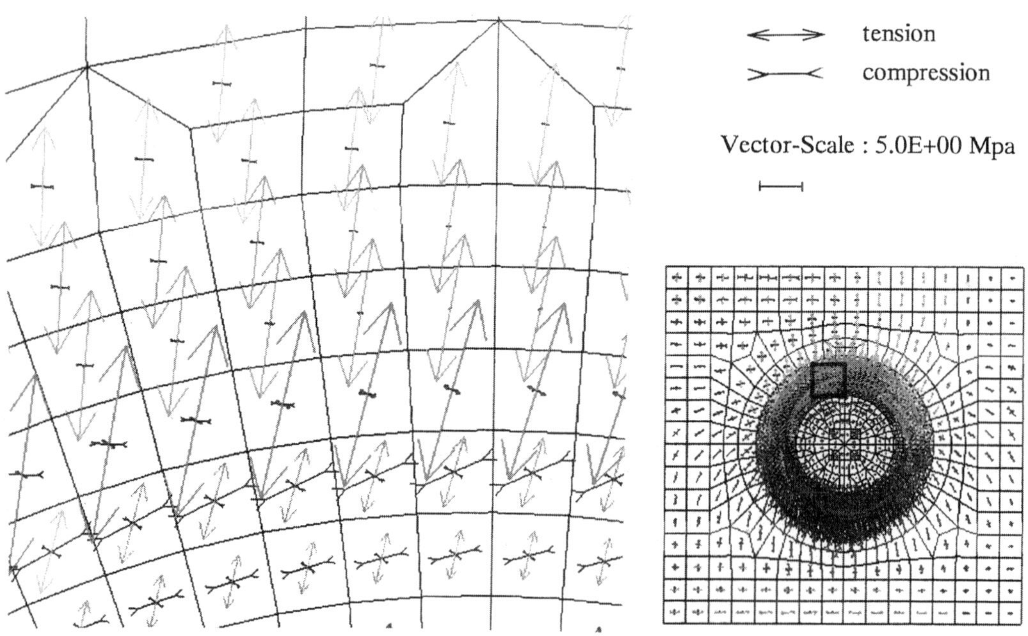

Fig.7 Interior stress condition of the spot welding (FEM)

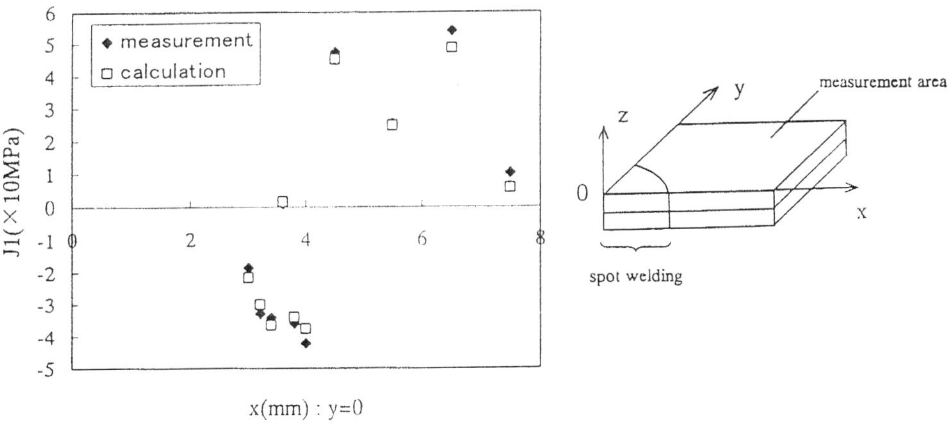

Fig.8 Comparison between the value of measurement J1 and calculation on the surface

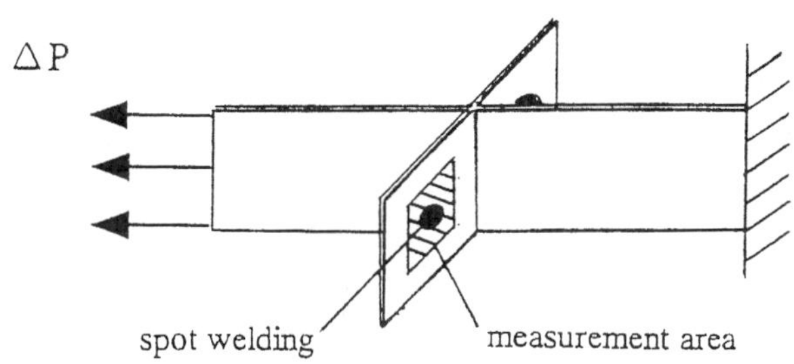

Fig.9 Experimental and analytical condition

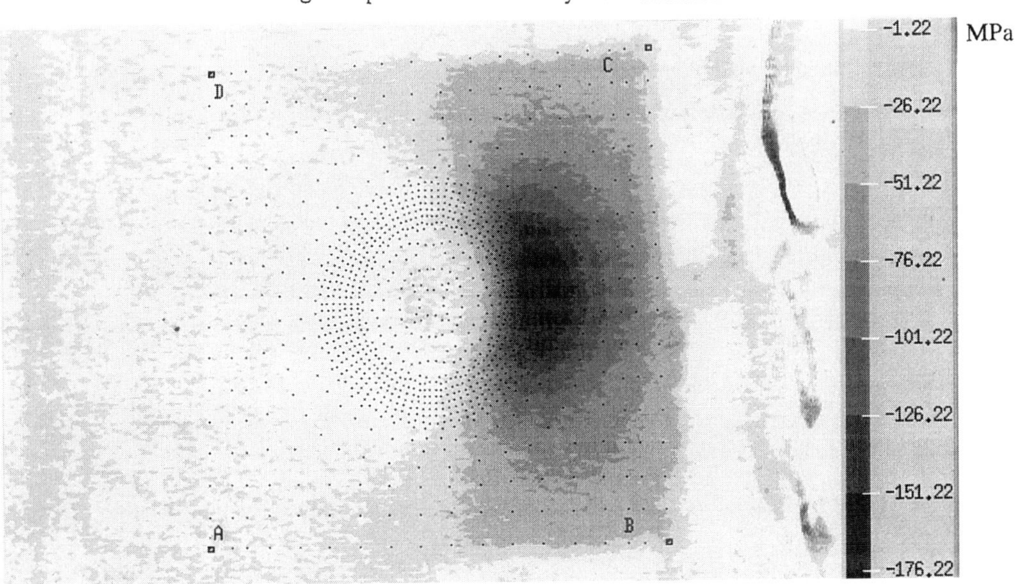

Fig.10 Thermoelastic image analysis on the part of the surface of Fig.9

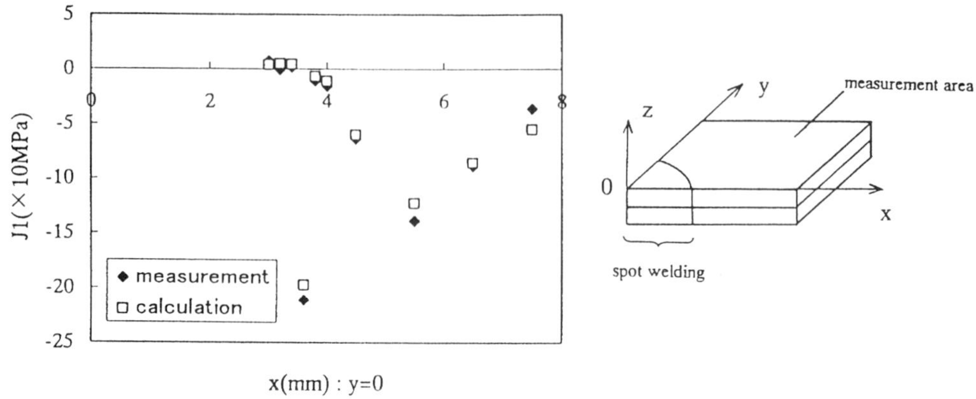

Fig.11 Comparison between the value of measurement J_1 and calculation on the surface

Fig.12 Testing situation of vehicle body part

Fig.13 Thermoelastic image analysis on the transport hook fixture

Fig.14 Thermoelastic image analysis on the spot welding of highest stress condition in Fig.13

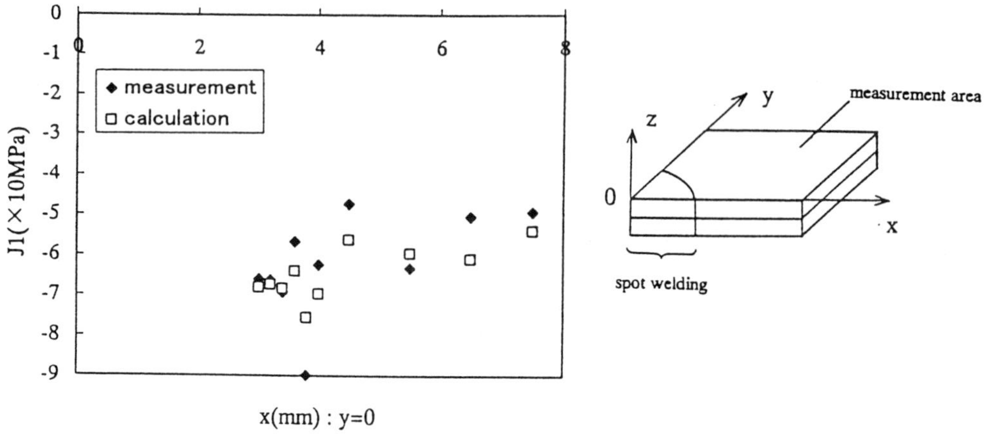

Fig.15 Comparison between the value of measurement J_1 and calculation on the surface

Step 1 Determination of analyzing points 【using Thermoelastic stress analyzer】

- Investigate the macro stress condition of the requested construction
- Pick up analyzing points in the requested construction
- Determine the analyzing area including that point
- Measure the micro stress condition on the surface of the analyzing area

↓

Step 2 Determination of boundary condition 【using the least square method】

- Construct the numerical model such as FEM of that analyzing area in detail
- Calculate the optimum boundary condition by comparing the measurement stress with the calculated stress

↓

Step 3 Calculation of the detail stress 【using finite element method】

- Calculate the principal stress value and directions
- Calculate interior stress conditions

Fig.16 Flowchart of analysis

REFERENCES

1. Murakami,Y., Yosimura,M.(1995) The Japan Society of Mechanical Engineers, 61-591
2. Yuuki,R.(1985) The Japan Society of Mechanical Engineers, 51-467
3. Compton, K.T. and Webster, D.B., Phys. Rev., Ser.2, 5, (1915), 159-166
4. Cummings, W.M. and Harwood, N.,(1985) Proc. SEM spring Conf. On Exp. Mechs., Las Vegas, Nevada., pp. 844-850

FATIGUE DESIGN OPTIMISATION OF WELDED BOX BEAMS SUBJECTED TO COMBINED BENDING AND TORSION

T. DAHLE
ABB Corporate Research, S-721 78 Västerås, Sweden

K-E. OLSSON
Volvo Articulated Haulers AB, S-351 83 Växjö, Sweden

J. SAMUELSSON
VCE Components AB, S-631 85 Eskilstuna, Sweden

ABSTRACT

Finding welding procedures to optimise design and life cycle cost of welded components is an important issue for industry today. This investigation tries to tie together modern production techniques and design tools in our study of longitudinally welded box beams with partial penetrating welds which were fatigue loaded in combined bending and torsion.

In the present investigation the use of a simple shear stress criterion to estimate allowable stresses as well as linear elastic fracture mechanics with fracture mode III to predict the fatigue strength as function of weld geometry is demonstrated.

KEYWORDS

Multi-axial fatigue, bending/torsion, mode III fracture mechanics

INTRODUCTION

Box beams are either manufactured by folding sheet material which is welded together in one single joint with a robot or by welding four sheets in four corners using backing strips and fully penetrating welds. By using folding and a welding robot one often runs into problems due to strict requirements on tolerances of the parts. Instead, as an alternative, it might be possible to weld in four corners without any backings strips or fully penetrating welds, which, hitherto, has been deterrent probably due to the initially existing crack-like root defects. This alternative is, however, not very much investigated.

Therefore, it is tempting both technically and economically, to try to answer some of the following questions, which also have been the aim of this research : for welded box beams subjected to a combination of bending and torsion loading, would it possible to obtain the optimum properties by applying single side welding without backing strips and/or fully penetrating welds by varying the weld geometry and manufacturing technology, is it possible to use a unified stress criterion to characterise their strength or fatigue life and/or is it meaningful to apply linear elastic fracture mechanics in design when a mixed-mode fracture mechanism is expected etc. ? For more details concerning the cost optimisation, refer to [3].

Optimum means highest bending-torsion fatigue strength at lowest production and service failure cost using the life cycle cost concept.

The aim of this work was further evaluation of work performed earlier in separate studies [1],[2] and [3] and to focus more on practical application of issues put forward in the above papers. The experimental part of the project consisted of testing 53 welded box beams in pure bending, pure torsion and combined bending and torsion, in phase (main part) and out of phase, using both median and high strength (with yield strength 350 and 900 MPa) weldable micro-alloyed steels. The work was accomplished within the latest Nordic co-operative project financed by the Nordic Industrial Fund, Nutek + Tekes and industry in Sweden and Finland.

FABRICATION, MATERIALS AND TESTING

Box beams of 150x150x2000 mm were fabricated in 350 MPa steel (thickness 8&10 mm) and even 900 MPa steel (thickness 6&8 mm) in a workshop with a welding robot as realistically as possible. Some beams were welded with transversal butt welds in the web with a backing strip. The beams and welds were left without any post treatment. Beam cross sections and testing rig are shown schematically on figures 1 and 2. In order to study the optimum welding, a number of beams were furnished with different weld geometries. Further details on weld geometries, steel compositions, strengths and testing details are documented in [1]. Refer also to figures 3 and 4 on weld geometry definitions. Testing, in summary, was carried out in an electro-hydraulic, computerised test rig with the ability to test the beams in either in-phase bending/torsion or out-of-phase bending/torsion. Only constant amplitude loading was applied during the tests.

APPLICATION OF A UNIFIED STRESS CRITERION

A detailed description of the results are given in [1]. The overall results from the experiments are shown here on figures 4 through 7. Figure 4 shows the results from pure bending and figure 5 from the pure torsion tests. Observe the difference in behaviour for the transversally joined beams compared to beams with longitudinal welds only. For all beams having transversal welds and tested in combined loading it was observed (as expected) that cracks grew along the transversal butt weld. Nearly all other beams had longitudinally running cracks, a few had a combination of both.

In [1] the hypothesis of a unified shear stress governed criterion to describe fracture for the combined loading was investigated. The results indicate the possibility, at least in an engineering sense, to use a simple theory proposed recently by Papadopoulos [4]. The Papadopoulos criterion can be written as follows :

$$\sigma_{eq} = \sqrt{\frac{\sigma_a^2}{3} + \tau_a^2} + \alpha \frac{\sigma_a + \sigma_m}{3} \leq \beta \qquad (1)$$

where σ_a and τ_a are the amplitudes of the normal stress due to bending and the shear stress due to torsion respectively and σ_m is the bending mean stress. The constants α and β are defined as

$$\alpha = \frac{t_{-1} - b_{-1} / \sqrt{3}}{b_{-1} / 3} \quad ; \quad \beta = t_{-1} \tag{2}$$

The evaluation showed that the fatigue data from the longitudinally welded beams did not follow the theory strictly. Figure 7 shows the data of combined loading unified with the criterion above. The straight line is the predicted criterion using the pure bending (b_{-1}) and pure torsion (t_{-1}) fatigue strength properties as input and applying a slope of -5. All cracks were observed to run longitudinally (in mode III) and showed no indication of bending stress influence (mode I). For the longitudinal welds the term α happened to be close zero or negative and was set to zero, whereas according to the theory α must be > 0. In any case, α is a relatively small number and, therefore, the error made is relatively small. The prediction seems to be acceptable for both types of welds, i.e. for both types of crack paths as can be seen in figure 7. Also seen are the results from the high strength steel (type 900) as well as for the 90° out-of-phase loading. There is not any significant effect of the phase shift in this case, a fact that is opposite to the results from Siljander (6). In figure 8 the experimental results from Sonsino [5], Siljander [6], and Yung [7] are re-assessed using the above criterion according to equation (1) and (2). The agreement is good, the error for the in-phase data is within a factor approximately 3. The conditions for the out-of-phase effects have to be investigated further. It can also be concluded that it is possible to benefit from using higher strength steel as shown in this case for the 900 steel.

A FRACTURE MECHNANICAL ASSESSMENT

Determination of Torsion Torque Strength as function of geometry

Using a stress criterion in design for determine allowable stresses are illustrated above. The scatter of the experimental results is hiding the weld geometry effects. If any, it should be easier to apply fracture mechanics which offers this possibility.

To demonstrate the relative effect of weld size and penetration on the fatigue strength of the beams the analysis was performed for torsion loading only. The effect of the welding process on these two parameters can start in the fabrication stage, e.g. by using non destructive testing to estimate penetration. If performed, it should also be possible to use the concept in design.

Here, it has been natural to determine the *Torsion Torque Strength* as function of weld size and penetration which in turn is used to optimise costs [2,3]. Hence, the intention is

- to determine stress intensity factor in mode III for a number of crack lengths using FEM
- to determine weld size and penetration together with life and slope from experiments
- which is used to estimate the *Torsion Torque Strength* and fatigue life as function of weld geometry

Refer back to figures 3-4 for the detailed weld geometry and crack paths which both are based on metallographic cross sectioning of a great number of welds.

Crack propagation (Paris' equation) in mode III can be written as

$$\frac{da}{dN} = A\Delta K_{III}^{n} \qquad (3)$$

The stress intensity can be normalised by torsion moment M,

$$\Delta K_{III}/M = f \qquad (4)$$

and from (3) and (4) :

$$\frac{1}{A}\frac{da}{f^{n}} = M^{n}dN \qquad (5)$$

Integrating

$$\frac{1}{A}\int\frac{da}{f^{n}} = \int M^{n}dN \qquad (6)$$

and designate the integral on the left side to ***I*** (="crack integral")

$$I = \int\frac{da}{f^{n}} \qquad (7)$$

The integral on the right side of eq. (6) is equal to the constant of the SN (Basquin: $S^nN = C$) curve

$$\int M^{n}dN = C \qquad (8)$$

where C is called "Capacity" [2] and defined as "strength in the time domain" which is life. . The slope of SN curve is assumed to be n=5 according to [1].

From (5),(6) and (7)

$$A = I / C \qquad (9)$$

The value of the crack integral, ***I*** , can be determined from FEM computations because f is a function of crack depth a. The crack integral, ***I*** , is consequently proportional to the Capacity-value C, which can be defined as a strength parameter in the "time domain". Therefore, $(\boldsymbol{I})^{1/n}$, is proportional to strength in the "force domain" as is the case for $(C)^{1/n}$. C and geometry can be determined from the experiments in the previous section.

Introducing, γ, in equation (6) as

$$\gamma = \frac{a}{L} \qquad (10)$$

which is the normalised crack depth, where L is the total crack path length (see fig.3 and 4), eq. (6) can instead be written as

$$\frac{L}{A}\int\frac{d\gamma}{f^{n}} = \int M^{n} dN \qquad (11)$$

and eq. (9) can be written as A

$$A = \frac{L \cdot I}{C} \qquad (12)$$

To solve the integral in eq. (6) a function, f, is introduced, which is assumed to be a monotonic increasing function of crack depth, a/L , with initial value f_0

$$f = f_0 + B \cdot \gamma^{b} \qquad (13)$$

This equation (or $f - f_0$) is fitted to a straight line in a log-log diagram by changing f_0. For each specific geometry the parameters B and b can be calculated.

From each tested beam a separate C-value (=M^n x N) is calculated which is transformed to a A-value according to eq. (12). A characteristic A_{50}-value, say at 50% probability, is determined assuming it to be a random variable following a Weibull distribution according to the method derived in [8]. The material constant, A_{50}, is used together with the crack integral in the following to estimate the fatigue strength which in the present case is the *Torsion Torque Strength* parameter designated M_{eq}. M_{eq}, derived below, is, accordingly, dependent upon the weld geometry, i.e. weld penetration and weld size.

The constant A_{50} (the mode III crack propagation constant) at stress ratio R = -1 was determined to be $A_{50} = 1.7 \cdot 10^{-14}$ (m, MPa) and was shown [2] to be comparable with literature data.

Written in mathematical form it can be written as

$$M_{eq} = \left(\frac{C}{N_{eq}}\right)^{1/n} \qquad (14)$$

with $N_{eq} = 2 \cdot 10^{6}$ cycles which is shown as predicted *Torsion Torque Strength* in figures 9 and 10 for the cases without weld preparation and with weld preparation respectively. It is seen that the two cases give fairly the same fatigue strength as expected.
Figures 11 and 12 illustrate the parametric fatigue life as function of weld penetration and weld size. From these figures it is clear that *penetration is twice as important as weld size.* A weld with weld size 5 and penetration 3.5 mm has the same life as a weld of size 7 and penetration 2.5 mm ! Add that the welding costs increase with the square of weld size, $(7/5)^2 = 2$, and whereas penetration is more or less ”free of charge”, makes it more worth while to

focus on means to increase weld penetration at reduced weld size. Another benefit is reduced welding distortion thanks to lower heat input with smaller weld.

Estimation of the weld throat thickness for shear stress calculation

The choice of weld throat thickness for shear stress calculation is natural in this case. Questions can though be risen as to its correctness. The results obtained above can be used for confirmation.

The predicted life can be used to calculate an equivalent weld throat thickness to the median fatigue strength value of 100 MPa which may be a reasonable choice according to the experimental results [1]. The ratio measured weld throat thickness divided by the calculated throat thickness will give us the weld throat thickness correction factor presented in figure 13. A factor less than one means that the throat value must be reduced.

For weld size 5 to 6 the weld throat appears to be an acceptable basis for shear stress calculation. This indicates that most of the stress values in (1) are acceptable. For large penetrations the weld throat value should be increased (maximum 25%). The equivalent weld throat thickness can be written as

$$h_{eq} = 3.23 + 0.436 \cdot (s + 2p) - 0.0244 \cdot s \cdot p \qquad (15)$$

where s = weld size and p = weld penetration, both in mm.

This confirms the dominating effect of penetration presented in figures 11-12. It is important to remember that eq. (15) is valid only for this box beam (8 and 10 mm thickness) and for the case without weld preparation and finally for a fatigue stress range of 100 MPa.

Mode III thickness effect

In figure 11 points indicate fatigue lives for weld size 4 for a decreased sheet thickness of 6 mm. The increase in life is as seen 1.4 corresponding to an increase in strength of $(1.4)^{1/5}$ = 1.07. One half of this increase is due to the increase in the enclosed section area of the box beam, as sheet thickness is reduced with constant outside dimensions.

The thickness ratio going from 8 to 6 mm is 8/6 = 1.333. That increase corresponds to 1.07 increase in strength which means a thickness exponent equal to z, i.e. $1.333^z = 1.07$ which leads to z=0.24. An objection could be that the relative weld size is larger for the 6 mm sheet. Still, this is very close to the recommended thickness exponent 0.25 in mode I cracking e.g. in Eurocode 3. Extrapolating the result for the 6 mm sheet to the same weld penetration as the 8 mm sheet from 3.3 to 4 mm for the same fatigue life ($2 \cdot 10^6$ cycles) this reduces the weld throat thickness from 4 to 3 mm which in turn would mean a welding cost reduction by 40%.

CONCLUSIONS

This investigation has shown that

- a simple shear stress criterion can be used to describe the overall fatigue strength of the welded beams from which it seems to be possible to used in design as allowable characteristic values when loading is constant amplitude, in-phase bending/torsion.
- the results did not show any significant effect of phase shift. The conditions for the phase-shift effect as seen in other investigations have to be investigated further.

- a clear beneficial effect of penetration over the weld throat thickness was stated with use of linear fracture mechanics and estimated mode III crack propagation parameters from the experimental results.
- a clear indication to use the above analyses in design is demonstrated.

ACKNOWLEDGEMENT

The first author wishes to acknowledge the co-authors for their co-operative support, patience and fruitful discussions during the preparation of this report.

REFERENCES

1 Dahle, T, Olsson, K-E, Jonsson, B : High strength welded box beams subjected to torsion and bending - test results and proposed design criteria for torsion/bending interaction, First North European Engineering and Science Conference Series (NESCO I), Welded High-Strength Steel Structures, Stockholm, Sweden, 8-9 Oct 1997, pp 143-161.

2 Olsson,K-E, Holm, D, Jakopovic, D : High strength welded box beams subjected to torsion and bending fatigue loads - mode I and III stress intensity factors and crack growth predictions, Ibid, pp 179-197.

3 Olsson, K-E : High strength welded box beams subjected to torsion and bending fatigue loads - optimum weld design considering welding cost and fatigue induced field failure cost, Ibid, pp 199-207.

4 Papadopoulos, I V : A new criterion of fatigue strength for out-of-phase bending and torsion of hard metals, Int. J. of Fatigue, Vol. 16, 1994, pp 377-384.

5 Sonsino, C M : Schwingfestigkeit von geschweissten Komponenten unter komplexen elasto-plastischen, mehrachsigen Verformungen, LBF-Nr 6078, 1993

6 Siljander, A, Kurath, P, Lawrence, F V, Jr : Proportional and non-proportional multi-axial fatigue tube-to-plate weldments, University of Illinois at Urbana-Champaign, Urbana, Illinois, Report to the Welding Research Council, 1989

7 Yung, J-Y, Lawrence, F V Jr : Predicting fatigue life of welds under combined bending and torsion, In: Biaxial and Multiaxial Fatigue, EGF 3, Ed. M W Brown and K J Miller, Mechanical Engineering Publications, London (1989), pp 53-69

8 K E Olsson : Weibull analysis of fatigue test data and the influence of scatter on the prediction error, Proc. of the conference Fatigue under Spectrum Loading and in Corrosive Environments, Lyngby, Denmark 26-27 Aug. 1993, pp 187-203

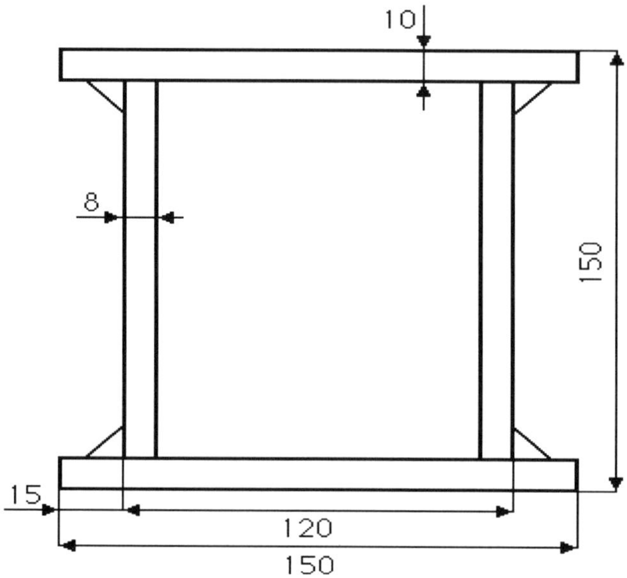

Figure 1 Cross-section of welded beams. From [1].

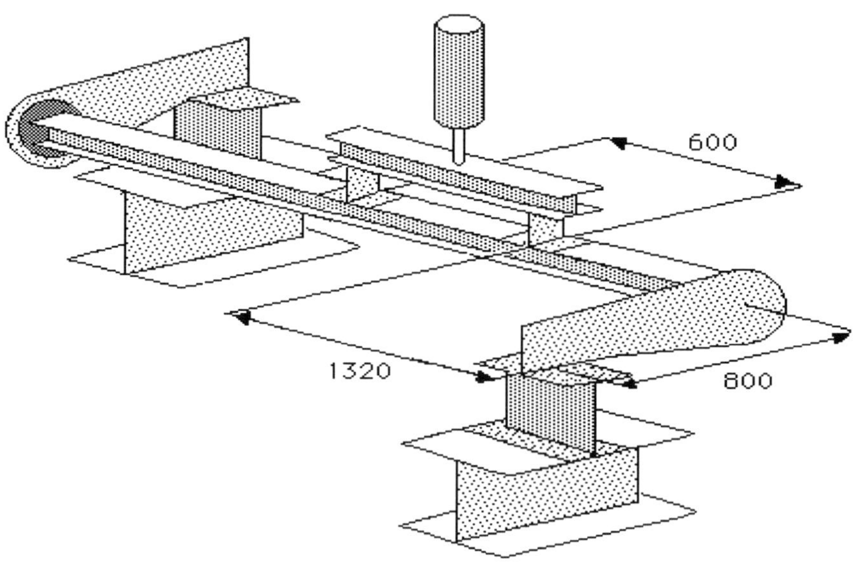

Figure 2 Testing rig loading principles for combined in-phase testing. From [1].

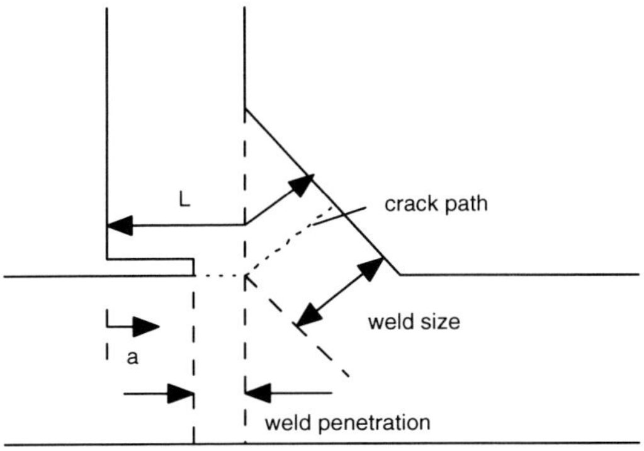

Figure 3 Model of crack path and weld geometry for weld without weld preparation. From [2].

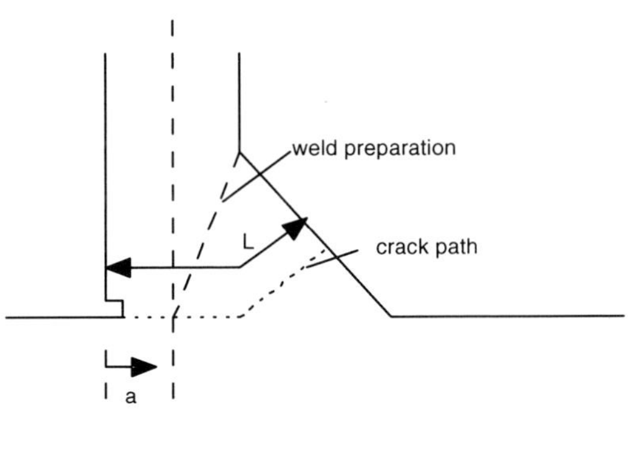

Figure 4 Model of crack path and weld geometry for weld with weld preparation. From [2].

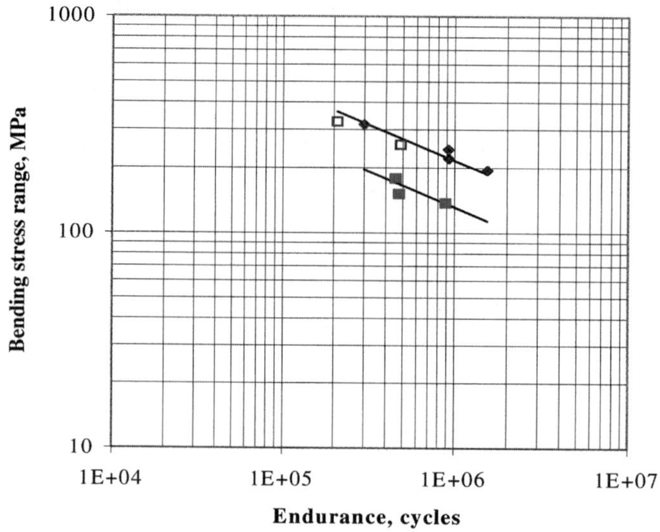

Fig. 5 Results from pure bending loading from beams with longitudinal welds (diamonds) and transversal welds (squares). Open rectangles are 900 steel. From [1].

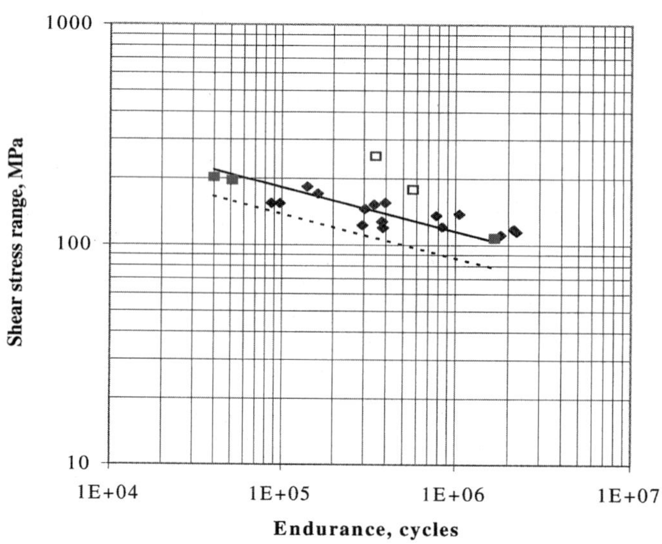

Fig. 6 Results from pure torsion loading on beams with longitudinal (diamonds) and transversal welds (squares). Open rectangles 900 steel. From [1].

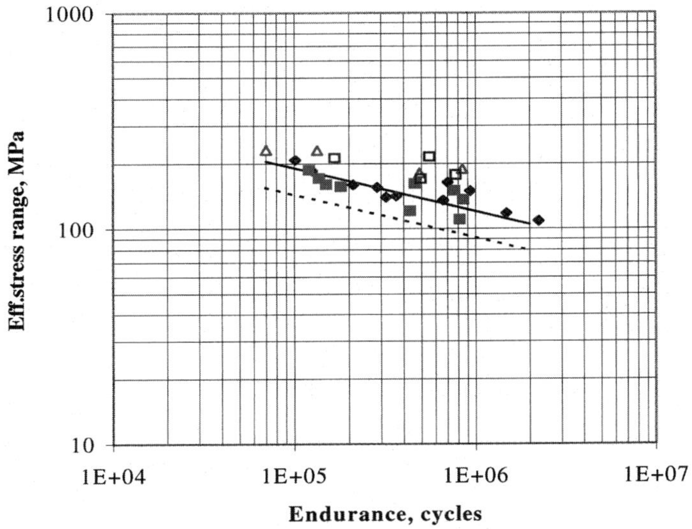

Figure 7 Data evaluated according to Papadopoulos model [4]. Diamonds are longitudinal welds, squares are transversal welds, open triangles with phase shift 90 degrees, open rectangles 900 steel. From [1].

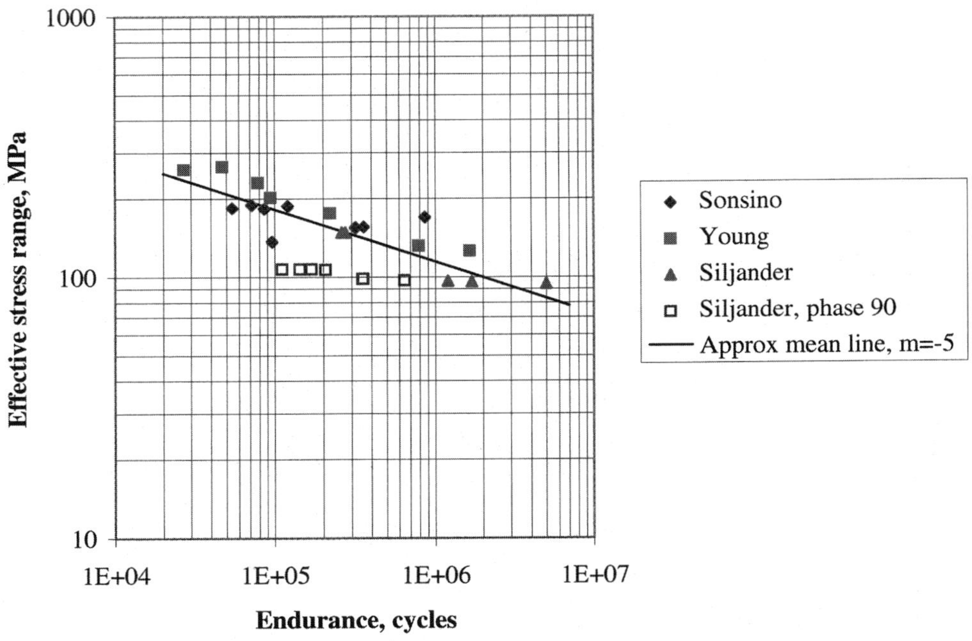

Figure 8 Data from Sonsino [5], Siljander [6] and Yung [7],on tube-to-plate joint re-assessed according to the Papadopoulos [4] criterion. Open square symbols from Siljander's out-of-phase results.

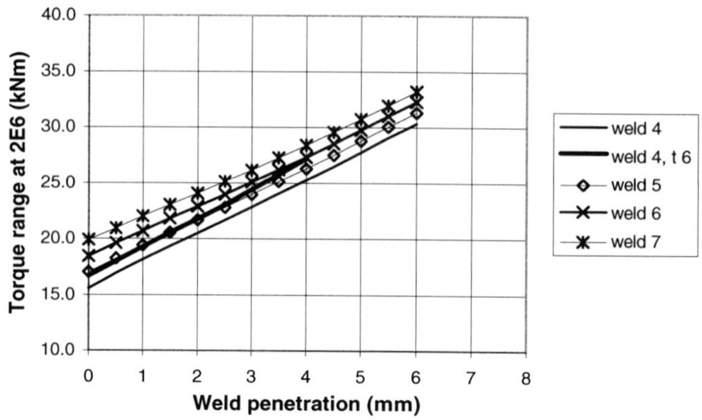

Figure 9 Predicted torsion torque range (R=-1) vs. weld size and weld penetration for welds <u>without</u> weld preparation. From [3].

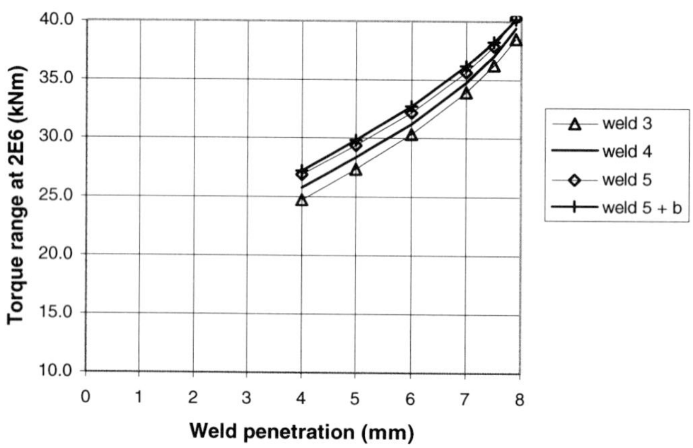

Figure 10 Predicted torsion torque range (R=-1) vs. weld size and weld penetration for weld <u>with</u> preparation (b means backing). From [3].

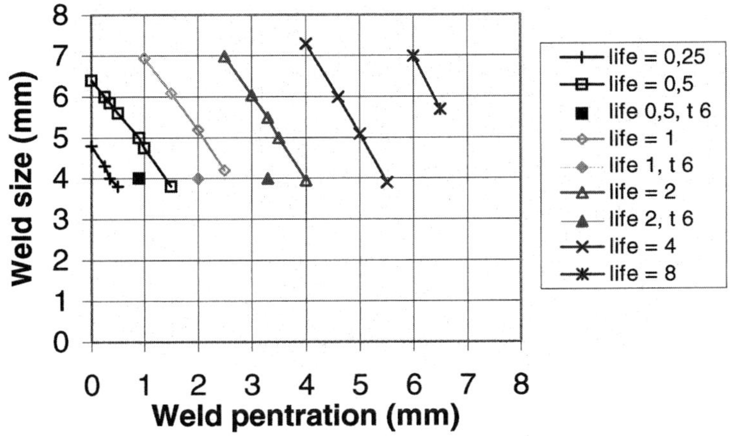

Figure 11 Predicted fatigue life vs. weld penetration and weld size for the case without weld preparation. From [3]

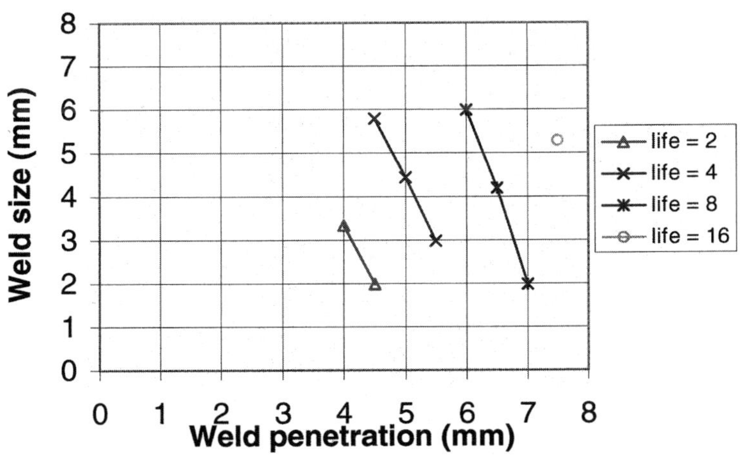

Figure 12 Torsion fatigue life vs. weld preparation and weld size for the case with weld preparation.

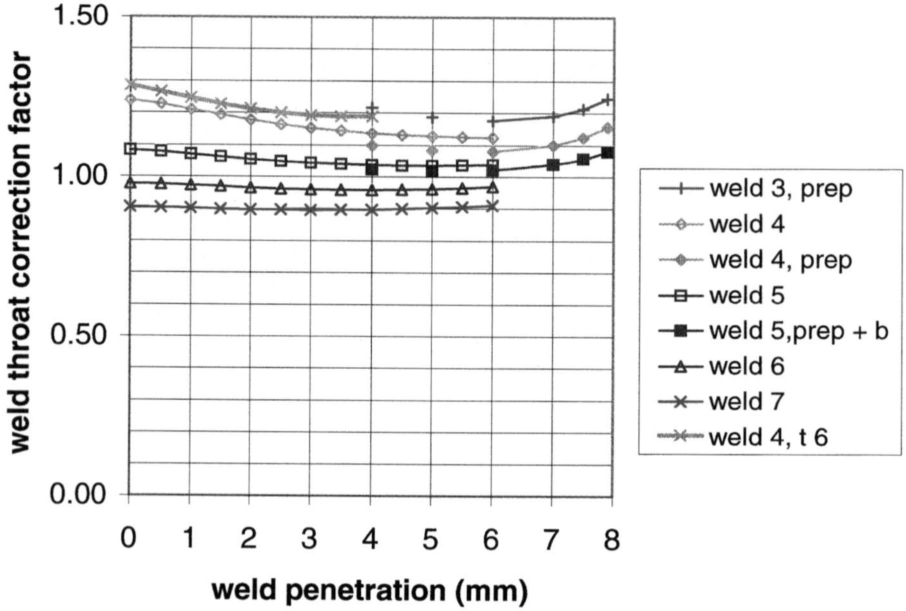

Figure 13 Weld size correction factor provided a fatigue shear stress range of 100 MPa at $2 \cdot 10^6$ cycles according to [1].

WELDING AND TIG-DRESSING INDUCED RESIDUAL STRESSES-RELAXATION AND INFLUENCE ON FATIGUE STRENGTH OF SPECTRUM LOADED WELDMENTS

L. LOPEZ MARTINEZ, R. LIN PENG*, A. F. BLOM** and D. Q. WANG*

Application Research and Development, SSAB Oxelösund AB, S-613 80 Oxelösund, Sweden
*Studsvik Neutron Research Laboratory, S-611 82 Nyköping, Sweden
**Aeronautical Research Institute of Sweden, P.O. Box 11021, S-161 11 Bromma, Sweden

ABSTRACT

Relaxation of residual stresses by spectrum fatigue loading and their influence on fatigue life have been studied numerically and experimentally for steel weldments. The experiments include spectrum fatigue testing and residual stress measurements by X-ray and neutron diffraction. The numerical studies include detailed stress analysis and modelling of fatigue crack growth.

The correlation between experimentally obtained fatigue life and fatigue crack growth calculations are good. This can be attributed to the accurate residual stress information through the thickness, provided by neutron diffraction measurements and accurate initial defect size assumptions used in the calculation.

Measurements of residual stress distribution after fatigue loading to certain numbers of cycles show that most of the relaxation of residual stresses occurred within 10% of the total fatigue life obtained with the load spectrum used in the study.

KEYWORDS

Fatigue, variable amplitude, steel, welding residual stress, relaxation, neutron diffraction, FEM analysis

1. INTRODUCTION

Along with stress concentrations and weld defects the residual stresses are one of the determinant parameters controlling the fatigue strength of welded joints [1]. In normal fatigue design, the level of residual stresses is often unknown and therefore these are assumed to reach the yield strength of filler metal. However, parameters in a welding process such as the welding technique used and the heat input should have influence on residual stresses, [2]. Furthermore, the variation of residual stresses through specimen thickness can be as important as the maximum level of the residual stresses since it influences the rate of crack growth previous to final failure. These aspects are even of greater importance when fatigue life improvement techniques are applied at weld toe region. Such is especially the case when the fatigue life improvement technique is based on the compressive residual stresses and the effectiveness of these methods is depending on the degree of relaxation under service. As a result in those particular cases the improvement can not be taken into account under the entire service life. In the case of TIG-dressing, the technique utilised in this study, the stress concentration at the weld toe is reduced as well as the residual stress field.

As the change of residual stresses through thickness by relaxation can influence the fatigue crack growth in welded components, it is of paramount importance in modelling fracture mechanics crack growth to know the residual stress distribution through thickness.

The present paper deals with welding residual stress relaxation due to static and fatigue spectrum loading on both as-welded and TIG dressed specimens.

2. MATERIALS AND FATIGUE TESTING

The material used in this investigation has the mechanical properties and chemical composition shown in Table 1 and Table 2, respectively. For further details, see [1].

Table 1. The mechanical properties

Yield strength (MPa)	Tensile strength (MPa)	Elongation, A5 (%)
615	747	31%

Table 2. The chemical composition

C	Si	Mn	P	S	Al	Nb
0.09	0.21	1.63	0.11	0.02	0.03	0.024

The geometry of the specimens is shown in Fig.1. They were also used in previous investigations [1] and that give us a good reference about the fatigue behaviour of this specimen. Both a single static load and variable amplitude fatigue testing were carried out.

The welding procedure was MAG with 1.6 mm electrode, current 185 Amp (DC), voltage 23.5 and heat input approximately 1.5 kJ/mm with consumable PZ 6130 (Mison 25) without pre-heating. The welds on the sides of the stiffeners as well as at the corners has been produced in an alternating diagonal sequence in order to limit the interpass temperature (<250° C).

Since all the specimens have been produced in the same way, the residual stress field induced by the above described weld procedure should be very similar in shape. By that the measurements carried out here should be representative for the rest of the specimens tested within the Nordic programme for the same steel [10].

Fatigue testing includes constant amplitude and spectrum loading. The spectrum is a randomised sequence created within each block of 500000 cycles by employing a draw without replacement routine. The blocks were repeated without reseed until fracture occurred. This gives an entirely randomised sequence until failure. The spectrum used, SP2 (R=0-0.77), is a straight line range-pair counted spectrum with I=0.99, see Fig. 2. The spectrum tests were carried out at FFA. The σ_{max} in the tested spectrum had been taken as the same value as for the static load applied in order to produce and document relaxation sequences.

3. RESIDUAL STRESS MEASUREMENTS BY DIFFRACTION TECHNIQUES

Surface and through thickness distributions of residual stress were obtained by X-ray and neutron diffraction measurements, respectively.

Neutron diffraction measurements

The neutron diffraction measurements were carried out with the REST diffractometer at the R2 reactor in Studsvik. The neutron wavelength was 1.76 Å. The technique uses atomic interplanar spacing as internal strain gauge and measures changes in the interplanar spacing due to stresses. Residual stress distributions at three different cross sections were investigated, see Fig. 2 and 3. They were chosen partially according to the fatigue behaviour of the specimens, partially for a more complete picture of the welding stress distribution. The first cross section, indicated by "A", was chosen to be near the weld toe and was 13 mm from the end of the attachment. The second cross section, named "B", were found to be prone to fatigue failure for the TIG-dressed specimen. The last cross section "C" was at the mid-width of the plate. The specimen's natural co-ordinates, i.e. the longitudinal, transverse and normal directions, were assumed to be the principle directions of stress. Lattice strains along these directions were obtained from measurement on also Fe <211> reflection:

$$\varepsilon = \frac{d - d_o}{d_o} \tag{1}$$

where d and d_o are the stressed and stress-free interplanar spacing, respectively. Residual stresses were then derived using Hooke´s law:

$$\sigma_i = \frac{E}{1+\nu}\left(\varepsilon_i + \frac{\nu}{1-2\nu}\sum_j \varepsilon_j\right) \tag{2}$$

with Young's modulus, E=225 MPa and Poisson's ratio, ν=0.284, respectively.

The specimen geometry indicates a symmetrical stress field around the mid-plane and around the mid-width, which were confirmed by preliminary neutron diffraction measurements and by X-ray diffraction measurements at the surface. Therefore only through half-thickness stress distributions were mapped. The incident slit, which defined the size of the incoming neutron beam, was 2 mm wide and 2 mm high. With a receiving slit of 2 mm wide, spatial resolutions in all the three directions can be approximated to 2 mm.

The stress-free lattice spacing were obtained by measuring on small coupons cut from different locations in an as-welded plate and an as-welded and TIG-dressed plate. They were cross checked by measuring in each specimen at a location which was far away from the weld. Standard deviations in strains were typically smaller than $\pm 1\text{x}10^{-4}$, calculated from uncertainties in peak fitting. The corresponding errors in stresses are less than ±25 MPa.

One specimen for each condition, namely as-welded and TIG-dressed, respectively, has been measured before and after being statically loaded up to 250 MPa. One specimen for each condition, as-welded and TIG-dressed, has been measured before fatigue load and after 100.000, 500.000 and 2.000.000 cycles. In all these measurements all the stress components have been documented. Since the measurements have been carried out at three different locations the amount of results is very large so we are going to concentrate on only the longitudinal stress distributions in Location A, see Fig. 2.

Residual stresses obtained by neutron diffraction

Tensile residual stresses due to welding were observed in both A and B sections in the as-welded specimen, se Fig. 5 and Fig. 6. The maximum stress, close to yield strength of the material, was found near the surface at the weld toe. It decreases with increasing distance from the weld toe and from the surface. The application of TIG-dressing on the weld toe has a strong effect on the local stress distribution. As is shown by comparing Fig. 5 and Fig. 3, the tensile stress peak was shifted from near surface to subsurface and the maximum stress was increased from 556 to 699 MPa. As a result, much lower tensile stress was found near the surface. This is consistent with X-ray diffraction measurements at the surface where tensile stress was decreased from 360 to 256 MPa by the TIG-dressing operation. These results confirm the hypotheses that TIG-dressing increases fatigue resistance not only by improving weld geometry, i.e. reducing stress concentration factor, but also by reducing the tensile residual stress near the surface. At the TIG-dressed edge, the tensile stress was increased near the surface while the compressive stress near the specimen edge became larger, compare Fig. 8 and Fig. 6.

X-ray diffraction measurements

The X-ray diffraction measurements were carried out at Linköping University. Part of these results have been reported in [4]. Longitudinal stress at the mid-width of the plate and near the weld toe, have been measured. See C and A sections, in Fig.3. The measurements have been carried out after removing a surface layer of about 0.1 to 0.2 mm by electrolytic polishing. The irradiated area was approximately 4 mm to 6 mm. Cr-K_a radiation was used to measure the Fe <211> reflection at 5 ψ-angles. Residual stress was then determined from the slope of d<211> versus $\sin^2\psi$ distributions with an elastic X-ray constant for 211 plane. The elastic X-ray constant is defined as follows: $E/(1+\nu)=174.097$ GPa where $E=224$ GPa and $\nu=0.29$.

Comparison of residual stress measurements by X-ray and Neutron diffraction

In Fig. 9 we present a comparison of measurement results at near surface by neutron diffraction at 1 mm depth and X-Ray at about 0.1 mm depth for the as-welded specimen at the weld toe region, A location. To document the possible influence of TIG-dressing procedure we include the same comparison for a specimen in the TIG-dressed condition, see Fig. 10.

4. NUMERICAL MODELLING

Fatigue crack propagation

The numerical model used to predict fatigue crack propagation is a strip yield model based on Dugdale-Barenblatt assumptions but extended to leave plastically deformed material in the wake of the extending crack tip due to both fatigue crack growth and the weld induced residual stresses. This model was previously developed and was shown to be applicable both for plane stress and plane strain conditions by incorporating a variable constraint factor [5]. A constraint factor $\alpha = 2$ was used for the material, in our case steel, for the crack growth analysis model, based on comparison with elastic-plastic FEM calculations [6], to account for the three dimensional effect at the crack tip essentially leading to plain strain conditions.

The model was applied to both the as-welded and the TIG dressed condition. The influence of the residual stress fields on fatigue crack growth is accounted for by a concept of residual stress intensity factor. Such stress intensity factors are determined by the residual stress distributions at the crack site using a 3D weight function method [7] based on the residual stress in the crack growth planes. In the analysis of the fatigue crack growth, the residual stress intensity factor represents the influence of residual stress fields on the crack growth quantitatively, and will be added to the stress intensity factors caused by the cyclic loading. The redistribution of residual stress fields is accounted for by the procedure of calculating the residual stress intensity factors using the superposition principle of linear elasticity under elastic consideration [8]. Crack tip plastic deformation under both applied load and residual stress is accounted for in the elastic-plastic crack growth analysis model.

Elastic Stress Distributions

Finite element 3D solid models have been created both for the as-welded and the TIG-dressed specimen to analyse the stress distributions in the weld toe region. 20-noded isoparametric brick elements were used to achieve good accuracy in the stress results. One eighth of the specimen has been modelled due to the symmetry. Very small elements were created near the toe of the weld to account for the dramatic stress concentrations, especially for the as-welded specimens. The finite element models are shown in Fig. 11.

The weld toe radii were obtained from measurements on several specimens and average values of 0.14 mm obtained for the as-weld specimens, and 7.0 mm for the TIG-dressed specimens, respectively. The finite element analysis is based on linear elasticity. The computations reveal such high local stresses that plastic deformation will occur for most of the load levels applied in the testing, especially for the specimens in the as-welded condition.

Redistribution of Residual Stresses

A simple computation of residual stress relaxation can be made by assuming an elastic perfectly-plastic material constitutive behaviour with a flow stress of the average of the ultimate stress and yield stress, and that the plastic deformation changes only the local stress distribution. Together with the finite element analysis and the measurement of the initial residual stress, the relaxation of the residual stress can be computed based on the cyclic material behaviour shown on the left side of Fig. 12. The analytical result is shown as a solid line in the right hand side of Fig. 12. The analytical result shows a good agreement with the experimental result. The meaning of this comparison is very clear. The relaxation of residual stress under fatigue loading is mainly due to the plastic deformation under the peak fatigue load. Therefore, different stress levels may create different relaxation of residual stress for the same welded configuration.

5. RESULTS AND DISCUSSIONS

Results of fatigue testing

The results of fatigue testing are presented in Figures 13 and 14. Constant amplitude data include as-welded conditions (mill scale and blast cleaned parent plate surface) as well as TIG-dressed condition, always with mill scale. In Fig. 8 fatigue lives are plotted both versus maximum stress in the load spectrum and also versus equivalent stress, [8]. Spectrum fatigue test results show the beneficial effect of TIG-dressing. Also shown in Figs 8, 13 and 14 are Computed Model Predictions (designed CMP in the diagrams). These are further discussed below.

Fatigue Crack Propagation

With the numerical model described in section 4, fatigue crack growth is analysed with residual stress distributions in the initial condition for both the as-welded specimens and the TIG-dressed specimens, and the residual stress relaxation based on the simple plastic deformation consideration for different stress levels. The fatigue crack growth rate for DX 590 is expressed in tabulated form in Table 3 below.

Table 3. The crack growth rate for Domex 590

da/dN (m/cycle)	9E-10	6.6E-9	2.6E-8	1.3E-7	6.3E-7
ΔK_{eff}, MPa $\sqrt{m}$	4.30	9.10	14.30	21.13	35.50

The crack growth analysis is firstly performed for the constant amplitude loading, with a stress ratio of R=0, for both the as-welded and the TIG-dressed specimens. The computations were started with an initial flaw size of 0.15 mm in depth with an aspect ratio, a/c=1. This initial flaw size is the average of experimentally observed imperfections at weld toe in unloaded welded specimens. The analytical results are shown in Fig. 13 as piece-wise lines in the plot of experimental SN-curves.

Despite the compressive residual stress present on the surface at the weld toe after residual stress relaxation as shown in Fig. 14 for the as-welded specimens, their fatigue lives are still significantly lower than those for the TIG-dressed specimens, revealing the importance of the degree of stress concentration for fatigue strength. The fatigue life is mainly related to the range of stress, rather than the maximum value, at the stress concentration/crack start site. The stress range is not effectively reduced by the compressive residual stress which can only reduce the stress ratio. However, the crack growth rate seems not to be very much affected by the stress ratio. Therefore, the most effective way to increase the fatigue strength for welded joints seems to be a reduction of the stress concentration at the weld toe.

For more realistic situations involving both spectrum loading and various initial flaw size distributions, the fatigue life may vary significantly depending on both the load spectrum and initial flaw sizes. Analyses are therefore performed for the actual spectrum load and a range of relevant initial flaw sizes. One of these analyses is showed in Fig 15 for the computed fatigue life of the TIG-dressed specimen. The same peak stress level σ_{max}=250 MPa as for relaxation studies, has been analysed. The fatigue life is shown as function of both the initial flaw size in the range of 0.1 to 1mm, and the weld toe condition as-welded and TIG-dressed. The fatigue life for the TIG-dressed specimens show much better fatigue life than the as-welded specimens. However, TIG-dressed specimens are much more sensitive to the initial flaw size than as-welded specimens. This is mainly due to the fact that most of fatigue life is consumed within the smaller part of the crack size region, for moderate stress ranges. This must be taken into account when annealing is going to be applied in order to improve the fatigue strength of welded components.

Effect of Weld Toe Geometry

For the fatigue crack growth analysis, the most critical location is at the stress concentrations where the crack can be initiated under cyclic loading. The finite element stress analysis shows consistently good agreement between the highest stress concentration locations and the experimental crack initiation sites, see the insert shown in Fig 16. The finite element analysis shows that the local stress concentration for the crack is very high at weld toe, with a stress concentration factor of around 5.8. The stress concentration, however, decreases rapidly in depth, see the open square symbols shown in Fig 16. There is, however, a significant area on the surface of the specimen near the toe with the high stress concentration, indicating the possibility of multiple crack initiations in a relatively large area.

The stress concentration can be effectively reduced by the TIG-dressing technique as the finite element analysis results in Fig 11b shows. The stress concentration has been dramatically reduced from 5.8 to about 1.6 after the TIG-dressing treatment. In addition, the change of stress through the thickness becomes much more smooth, see the results shown in Fig 16b. The reduction in the stress concentration in the TIG-dressed specimen is mainly due to the change of configuration at the weld toe. A much larger radius on the fusion line along the weldments near the global stress concentration area is obtained after TIG-dressing.

Redistribution of Residual Stress under Static and Cyclic Loading

Residual stresses through the thickness of the specimens are taken from [4]. In Figure 17, the longitudinal through-the-thickness distribution is shown at section A, 13 mm away from the flange in the symmetrical plane. This location is the most likely place that a crack may be initiated at. The stress in this location is used in the fatigue crack growth analysis.

Within the range of accuracy for the neutron diffraction measurements, the results shown in Fig17 indicate that stress relaxation occurs more strongly near the surface than in the interior of the specimen. 500,000 cycles produces more stress relaxation than 200,000 cycles. For a depth of more than 3 mm, there is basically no difference in the residual stress for various load cycles and the initial condition. Therefore, the residual stress relaxation is a near surface phenomenon for the case of as-welded specimens. Since a significant part of the welded joints fatigue life is consumed in the small crack growth region for small stress ranges, the residual stress near the surface region should be analysed in detail.

As mentioned above, the relaxation of residual stress depends on the level of load. An example will be discussed based on the simple solution demonstrated in the results shown in Fig12 for the residual stress on the surface of the specimen. The peak stress level is 250 MPa which is the same stress level as used in the spectrum loading. Fig 18 shows the normal stress variation through the depth at the same position as in Fig 17. The residual stress on the surface is an estimated value equal to the yield stress of the material. When the specimen is loaded, a high stress region appears near the surface. Using a perfect elastic plastic deformation relation to approximate the material constitutive relation schematically, as shown in the left side of Fig 12, the residual stress relaxation following unloading can be computed. The result is shown in Fig 19.

In this figure, the analytical results show that there is a significant relaxation of residual stress near the surface. Even after one single peak load only, the residual stress on the surface becomes compressive with a level near the compressive yield stress. Such a dramatic change of residual stress is due to the high stress concentration for the as-welded specimen. The results also indicate inefficiency in using mild surface treatments such as blast-cleaning to improve the fatigue strength under spectrum loading since the induced compressive residual stress at the surface is not high enough compared to the tensile residual stresses already approaching the yield stress. The only beneficial effect of the blast-cleaning technique may be the improvement in the surface condition and the possible increase in the root radius at the weld toe which may somewhat reduce the local stress concentration at the toe. The analytical results compare favourably to the measurements for depths larger than 2 mm, confirming the reliability in the simple evaluation of the residual stress relaxation. The analytical results are used in the analysis of fatigue crack growth and fatigue life of the as-welded joints.

For TIG-dressed specimens the residual stress distributions are quite different from those of the as-welded specimen, [4]. In particular, the residual stress near the surface is rather low. There is almost no residual stress relaxation near the surface after either static load or fatigue loading up to 500,000 cycles.

Nevertheless, there is some extent of relaxation taking place in the region of about a quarter through the thickness due to the initial high residual stress in this region. The relaxation for the static load is still quite close to the relaxation of residual stress after 500,000 cycles, indicating that the relaxation of residual stress is still mainly due to the plastic deformation of the material. The plastic deformation now occurs beneath the surface of the specimen instead of on the surface like for the as-welded specimens, [4].

Using the finite element results shown in Fig 16 and the residual stress distribution for the initial condition, the longitudinal normal stress distribution is computed and shown in Fig 18. Under a load level of 250 MPa, the stress on the surface at the toe is about 400 MPa, which is lower than the yield stress. Therefore, there is no plastic deformation on the surface under this load. At a depth of between 2 mm to 4 mm, the normal stress level is higher than the yield stress under the load of 250 MPa, indicating that plastic deformation will occur in this region. The residual stress will consequently be considerably relaxed following unloading.

Again, the simple method discussed above is used to compute the stress relaxation and the numerical results are shown in Fig 20 together with the neutron diffraction measurements for both the static load and fatigue loading after different number of cycles. The measurements show that the residual stress relaxation is a gradual process. The longer the specimen is subjected to fatigue load, the more the residual stress will be relaxed. The relaxation due to the static load is very close to the stabilised residual stress state after 200,000 load cycles, Fig 19. The result from the simple model evaluation of residual stress is also quite close to the stabilised residual stress state.

The gradual relaxation of the residual stress may be explained by the rate of the application of the peak load. Under rapid loading, the material reaction is different from that of the slow static load, usually resulting in an increased yield stress. There needs to be many cycles of fatigue peak load before the material reaction is stabilised. The measured results show that the stabilised residual stress relaxation is rather close to the relaxation due to the static load. Very close estimation of the residual stress relaxation can be made even from the simple consideration of the plastic deformation due to the combined effect of applied stress and the initial residual stress distribution. Therefore, it is possible to analyse the fatigue crack growth in the joint under different load levels based on the finite element stress results and the initial residual stress results for both the as-welded and the TIG-dressed specimens.

6. CONCLUSIONS

Good agreement between numerical modelling and experimental data has been found.

Such agreement requires good knowledge of relevant initial flaw sizes for the actual weld process and any used post weld treatment.

Also, good understanding of the full three dimensional residual stress distributions, and their relaxation behaviour under spectrum loading, is required.

Numerical modelling can then be used to assess improvements in fatigue behaviour following post weld treatments, and under arbitrary load conditions.

Experiments verified the possibility to use high strength steel under spectrum fatigue loading, once relevant post weld treatment was applied.

ACKNOWLEDGEMENT

This work was financially supported by NI (Nordic Industrial Foundation), NUTEK (Swedish National Board for Industrial and Technical Development, SSAB, ABB and FFA. The authors are indebted to Mr. Bengt Wahlstenius (FFA) for performing the fatigue tests and Mr Tommy Linden (SSAB Oxelösund AB) for performing welding and TIG-dressing.

REFERENCES

1. Lopez Martinez, L. and Blom A.F. "Influence of life improvement techniques on different steel grades under fatigue loading", Fatigue Design of Components. Edited by G. Marquis and J. Solin. ESIS Publication 22, Elsevier Science Ltd, 1997.

2. Legatt, R. "Welding Residual Stresses", ICRS 5, June 16-18, 1997. Linköping, Sweden.

3. Bogren, J., Lopez Martinez, L. "Spectrum fatigue testing and residual stress measurements on non-load carrying fillet welded test specimens" Proceedings of the Nordic Conference on Fatigue. Edited by A.F. Blom, EMAS Publishers, West Midlands, England, 1993.

4. Lopez Martinez, L., Lin R., Wang D. And Blom A. F. "Investigation of Residual Stresses in As-welded and TIG-dressed Specimens Subjected to Static/Spectrum Loading". Proceedings of the North European Engineering and Science Conference Welded High-Strength steel Structures. Edited by A. F. Blom, EMAS Publishers, West Midlands, England, 1997.

5. Wang, G. S. and Blom, A. F., "A strip model for fatigue crack growth predictions under general load conditions", Engng. Fracture Mech., Vol. 40, No. 3, pp. 507-533, 1991.

6. Blom, A. F., Wang, G. S. and Chermahini, R. G., "Comparison of crack closure results obtained by 3D elastic-plastic FEM and modified Dugdale model, in localised Damage", Computational Mechanical Publications, Springerverlag, Berlin, 1990, Vol. 2, pp. 57-68.

7. Wang, G. S., "A generalised WF solution for mode I 2D part-elliptical cracks", Engng. Frac. Mech. Vol. 45, No. 2, pp. 177-208, 1993.

8. Wang, G. S. and Blom, A. F., "Fatigue crack propagation in residual stress fields", 6th Int. Conf. on Mech. Behaviour of Materials, Vol. 4, pp. 627-632, Pergamon Press, 1991.

9. Blom, A. F., "Spectrum fatigue behaviour of welded joints", Int. J. Fatigue, Vol. 17, No. 7, pp 485-491, 1995.

10. Welded High-Strength Steel Structures, (1997) Proceedings of the North European Engineering and Science Conference. Edited by A. F. Blom, EMAS Publishers, West Midlands, England.

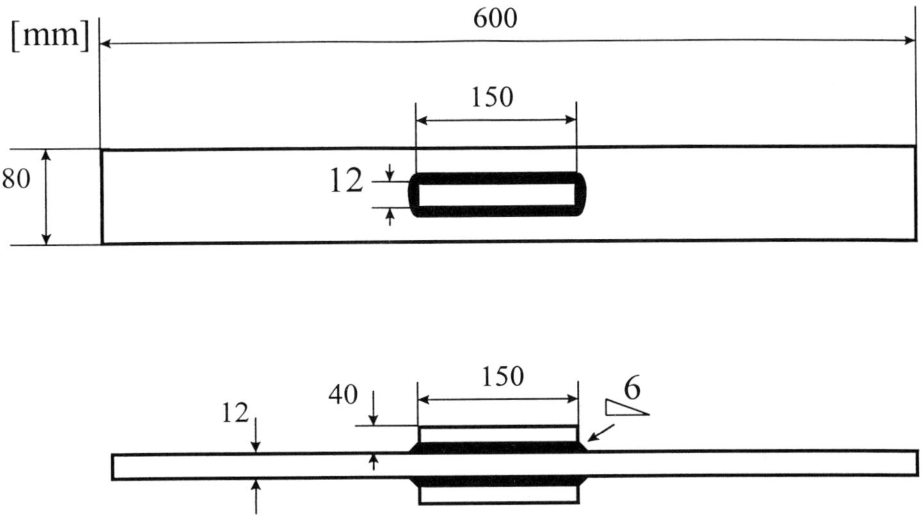

Figure 1 Fatigue test specimen

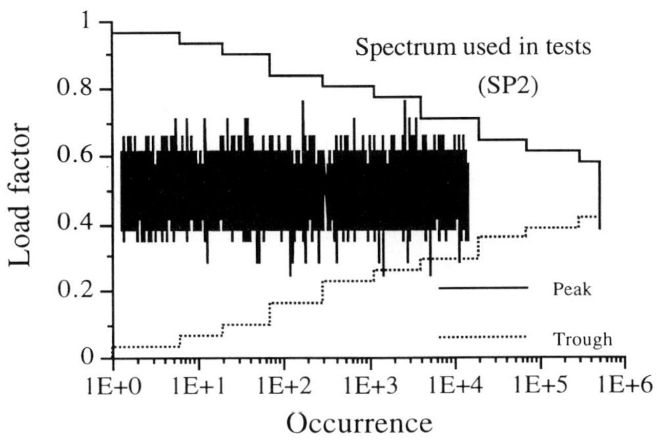

Figure 2 Load spectrum exceedance distribution

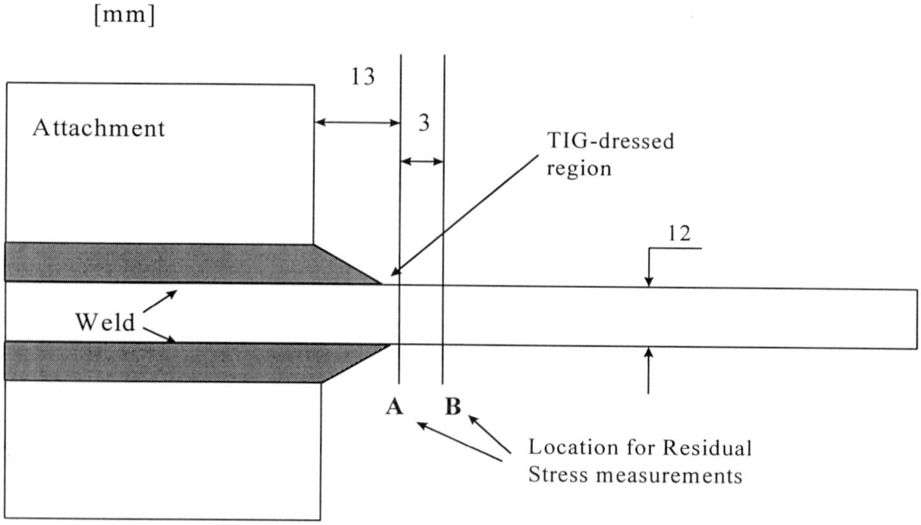

Figure 3 Specimen mid-section showing location for residual stress measurements

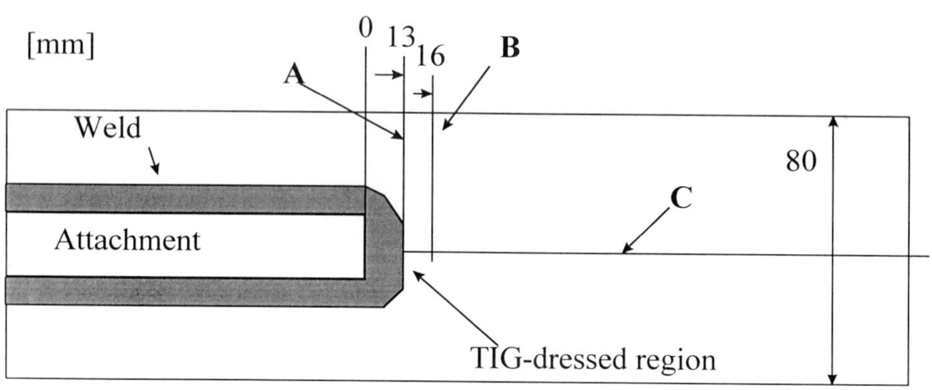

Figure 4 Upper view showing fatigue test specimen with locations for residual stress measurements

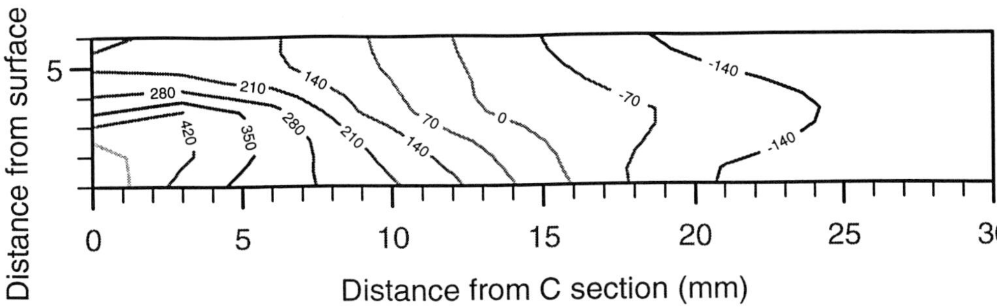

Figure 5 Longitudinal residual stress distribution in A-section of as-welded specimen

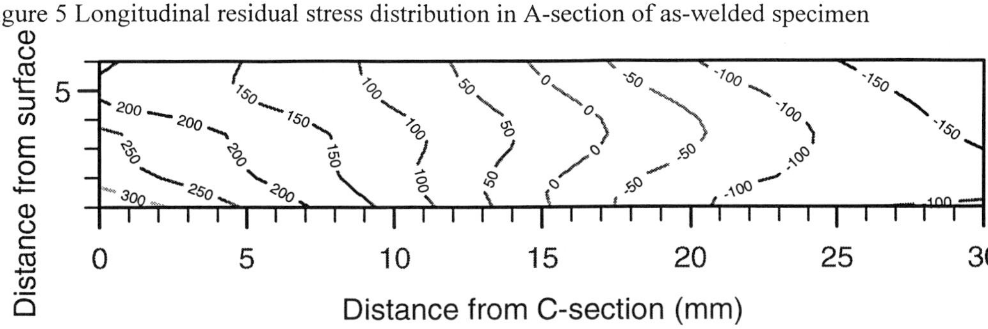

Figure 6 Longitudinal residual stress distribution in B-section of as-welded specimen

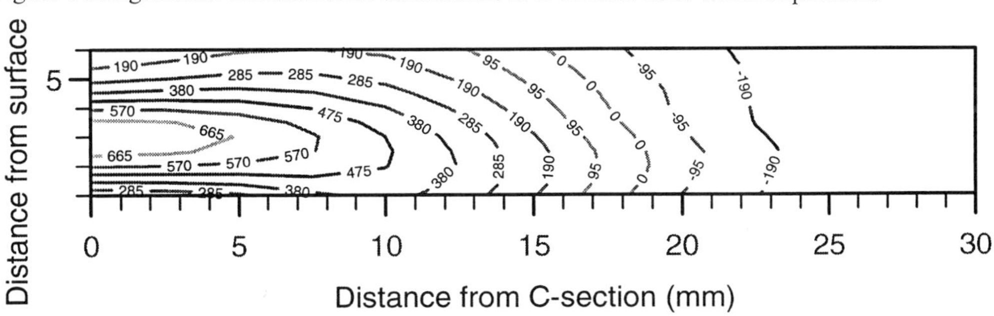

Figure 7 Longitudinal residual stress distribution in A-section of TIG-dressed specimen

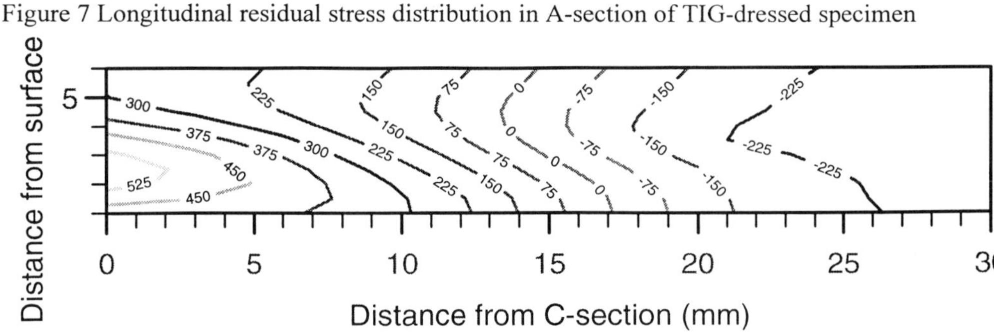

Figure 8 Longitudinal residual stress distribution in B-section of TIG-dressed specimen

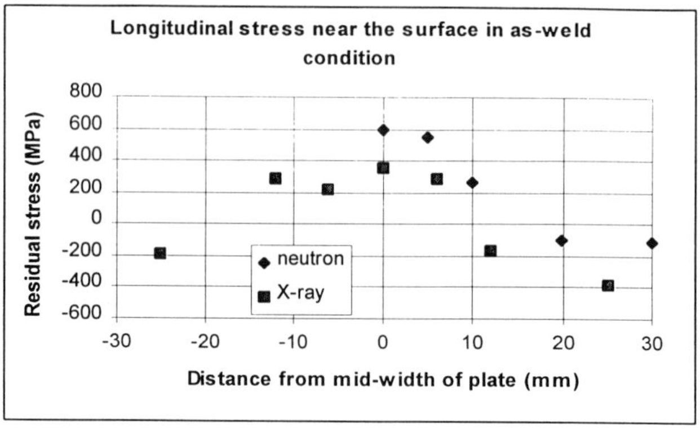

Figure 9 Comparison of measurement results at near surface by neutron diffraction at 1 mm depth and X-Ray at about 0.1 mm depth for the as-weld specimen.

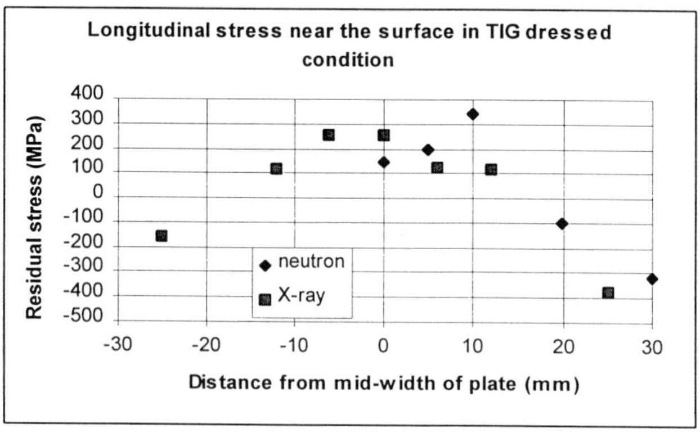

Figure 10 Comparison of measurement results at near surface by neutron diffraction at 1 mm depth and X-Ray at about 0.1 mm depth for the TIG-dressed condition.

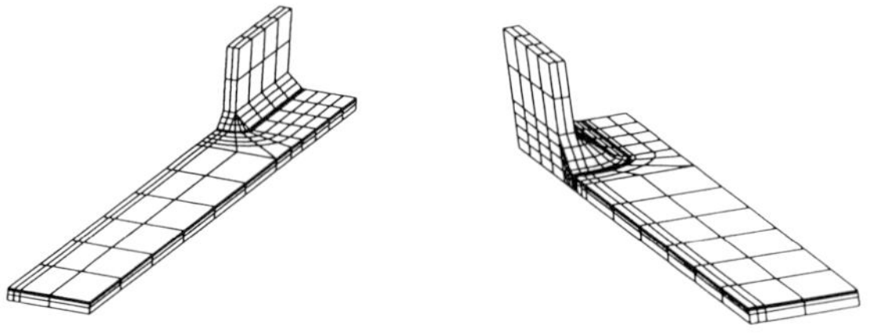

Figure 11 FEM model for as-welded and TIG-dressed specimens.

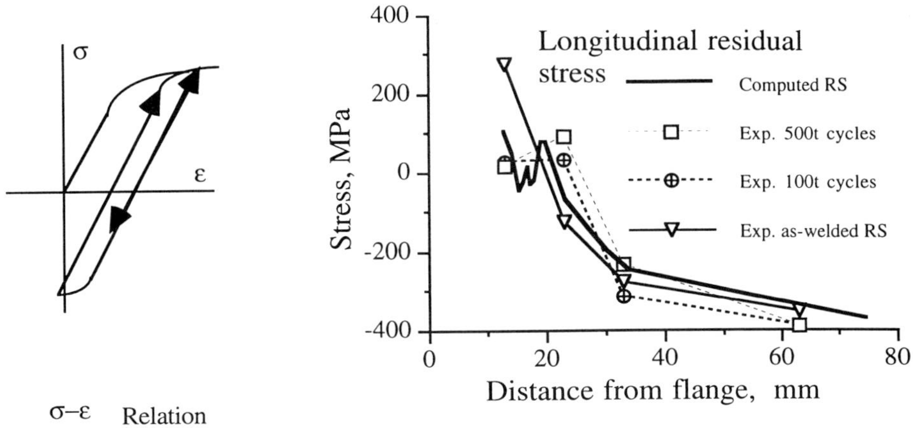

Figure 12 Residual longitudinal normal stress in CL ahead of attachment. Comparison of measured results and computed redistributed stress.

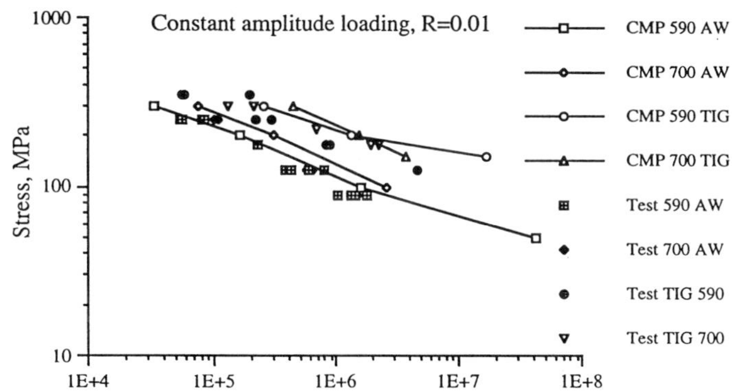

Figure 13 Constant amplitude fatigue test results and Computed Model Predictions (CMP)

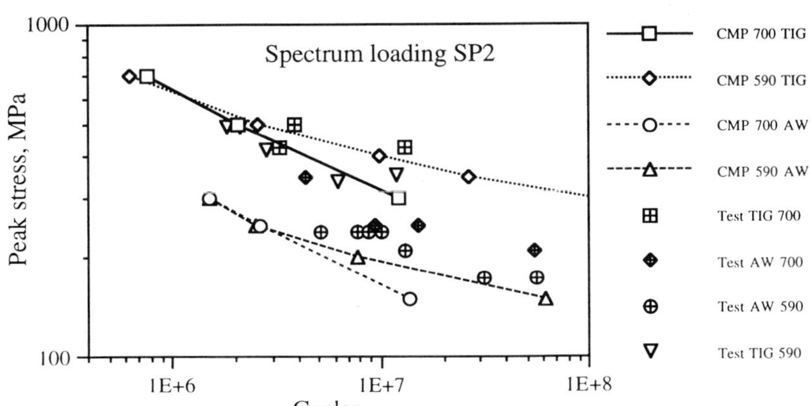

Figure 14 Spectrum fatigue test results and computed model predictions (CMP)

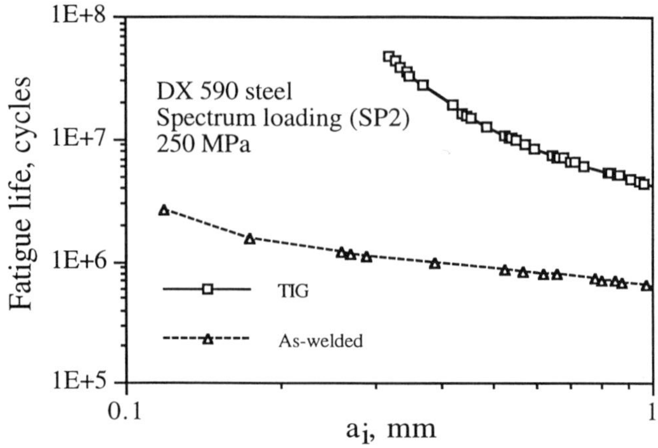

Figure 15 Fatigue life as a function of initial flaw sizes for as-welded and TIG-dressed specimens

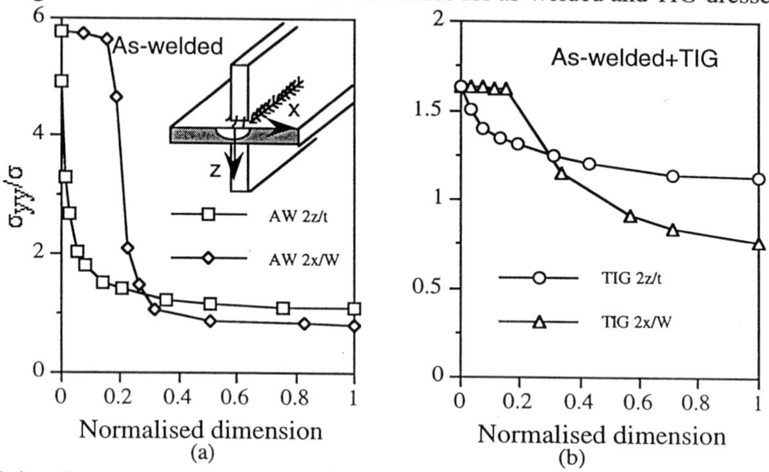

Figure 16 Finite element normal stress results at the crack growth plane

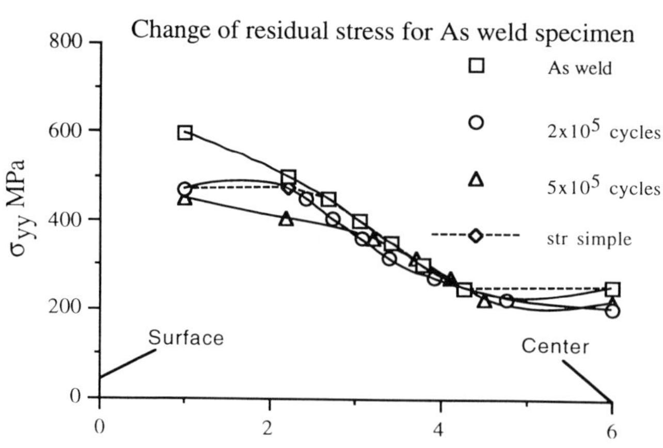

Figure 17 Through-the-thickness residual stresses for as-welded specimen

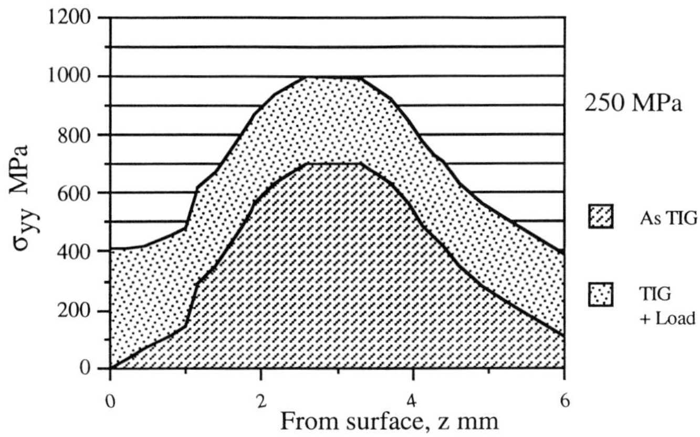

Figure 18 Normal stress variation through the depth before and after applied load causing plastic deformation in as-welded specimen.

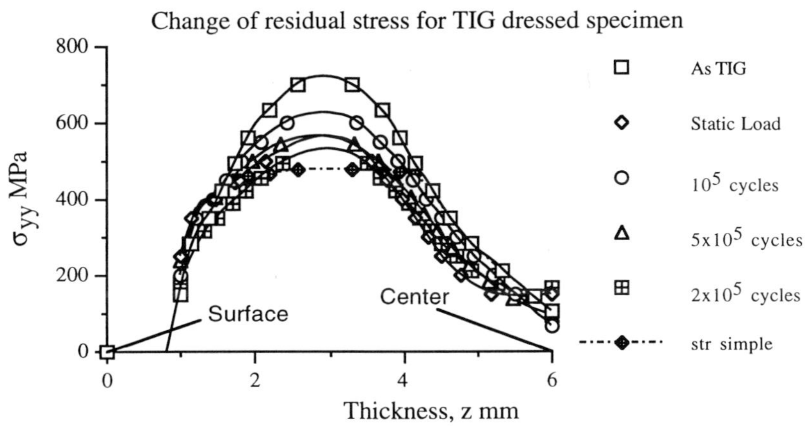

Figure 19 Comparison of through-the-thickness residual stresses for different conditions

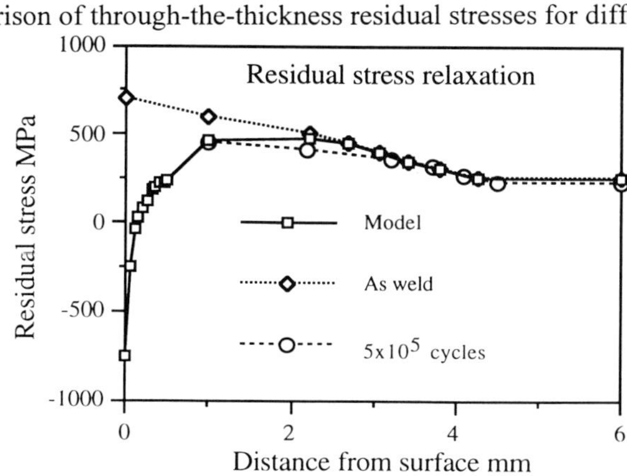

Figure 20 Relaxation of residual stress due to applied load

DATA ACQUISITION BY A SMALL PORTABLE STRAIN HISTOGRAM RECORDER (MINI-RAINFLOW CORDER) AND APPLICATION TO FATIGUE DESIGN OF CAR WHEELS

Y. MURAKAMI
Department of Mechanical Science & Engineering, Kyushu University, Fukuoka, 812-8581, Japan

K. MINEKI
Central Motor Wheel Co. Ltd., Anjo, Aichi, 446-0065, Japan

T. WAKAMATSU
Central Motor Wheel Co. Ltd., Toyota, Aichi, 471-0836, Japan

T. MORITA
Fukuoka Kiki Co. Ltd., Miyaki-gun, Saga, 841-0203, Japan

ABSTRACT

A very small strain histogram recorder based on the rainflow method has been developed and applied to strain measurement of car wheels under different road tests. Various strain amplitude histogram data under mountain road, city road and high-way were acquired by the recorder for various types of wheels. The data were evaluated by Miner's rule. The results of the damage evaluation have been used for the improvement of dimension and shape of automotive wheels.

KEYWORDS

Fatigue, Car Wheel, Data Acquisition, Small Strain Recorder, Rainflow Method

INTRODUCTION

Although car wheels are subject to complex loading in service, the fatigue design is not always based on necessary data obtained from real road tests due to the difficulty in strain measurement in service. Thus, the fatigue design is likely to be too conservative. However, car wheels are rotating components which strongly contribute to fuel consumption. It follows that in order to improve fuel consumption, a rational fatigue design method which enables one to reduce wheel weight must be established.

In this study, to measure the strain of car wheels in service, the authors used the portable strain histogram recorder (Mini-Rainflow Corder) which has been developed in a previous study [1] and has been revised in terms of noise resistance in the present study. The strain histogram measurements using the Mini-Rainflow Corder have been incorporated into the fatigue design procedure for car wheels to improve their fatigue reliability. The strain measurements were conducted by road tests on city road, mountain road and highway.

SPECIFICATIONS OF MINI-RAINFLOW CORDER

The newly developed recorder, Mini-Rainflow Corder, resolves strain waves based on the Rainflow cycle counting method [2,3] (Fig. 1) and stores strain ranges and cumulative frequencies into a small memory card [1]. The algorithm of the Rainflow cycle counting method was recently revised by Anzai and Endo [4,5] in a very simple form so that it can easily be installed in a IC chip [1].

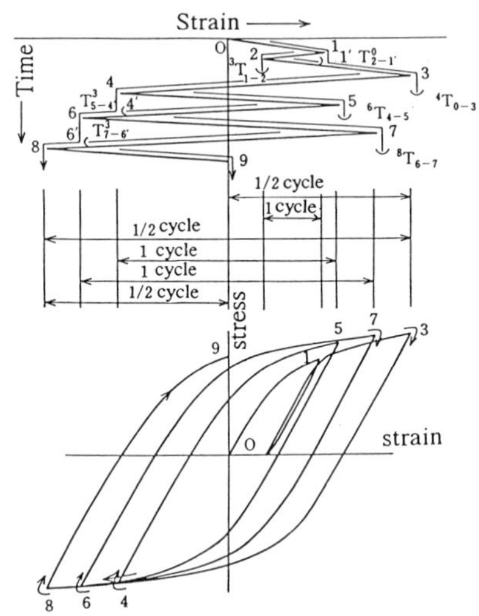

Fig. 1. The basic concept of the Rainflow Method [3].

The functions and specifications of Mini-Rainflow Corder are summarized as follows.

(1) The lead wire of a strain gauge can be easily and directly connected to the terminal of the recorder.
(2) The Rainflow algorithm is written in an IC chip in the recorder.
(3) The amplifier circuit is installed in the recorder.
(4) The bridge balance of strain gauges can be automatically performed.
(5) Resolution of strain waves up to 150Hz is possible.
(6) Resolved strain ranges are stored in a small standard memory card for personal computer (64KB SRAM card).
(7) Long time measurement using small conventional commercial batteries is available (greater than 10hr with 4 small 1.5V batteries).
(8) Measurement in the temperature range from −15℃ to 70℃ is available.
(9) Ability to endure accelerations up to 12G.
(10) Possible to attach to rotating machine components without using a slip ring.
(11) The minimum unit of strain measurement is 27×10^{-6} and the maximum measurable strain range is 7000×10^{-6}.

Figure 2 shows (a) the total view and the size of the recorder, and (b) the internal layout of the recorder.

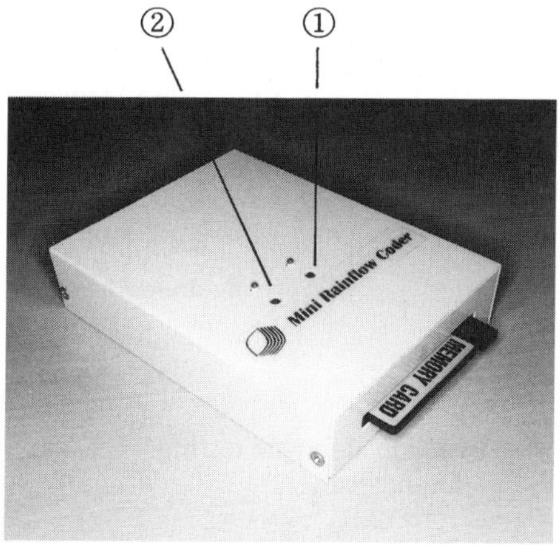

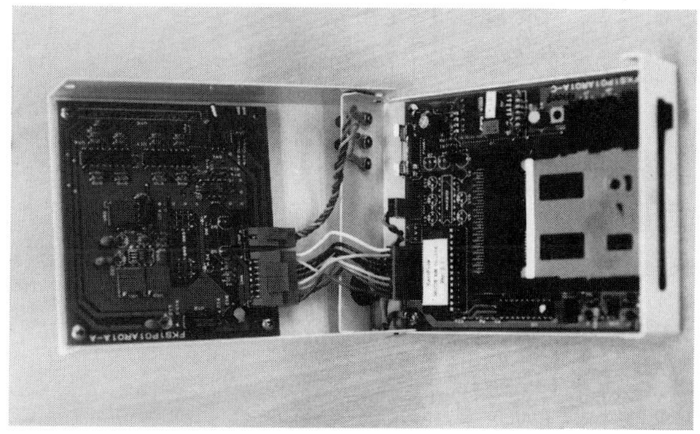

(a) The total view.
(Total size : 106mm×143mm×31mm).
(Weight : 321g)
① Bridge balance switch
② Start switch

(b) Layout of the circuits.

Fig. 2. The total view and the interior layout of the Mini Rainflow Corder.

APPLICATION TO CAR WHEELS

THE BASIC CONCEPT

Two conventional tests have been used to assure the fatigue strength of car wheels. One is the rotating bending fatigue test, so-called "cornering fatigue testing" (Fig. 3) and another is "drum testing or radial load fatigue testing" (Fig. 4). The strength of the part of "*hat*" is tested mainly by rotating bending fatigue testing. The strength of the rim, and the *weld* between the hat and rim, is tested by both rotating bending and drum testing.

The conventional fatigue design method for car wheels may be illustrated as Fig. 5. The bending moment (M) vs fatigue life (Nf) in rotating bending fatigue testing and the load (P) vs fatigue life (Nf) in drum testing are experimentally investigated. In this conventional design method, the correlation between laboratory test results and the performance of wheels in service is empirically based on past experiences from the performances of car wheels produced by a similar procedure. The performance of a new product is feedbacked to the fatigue design to modify or revise the model. However, this method cannot avoid delay due to this feedback time. But more importantly the degree of modification of the next model is not always clear, because the background of fatigue failure of the original model and its failure frequency in the market are not obvious. Thus, the design engineer is likely to introduce a large safety factor and accordingly the fatigue design of car wheels becomes conservative.

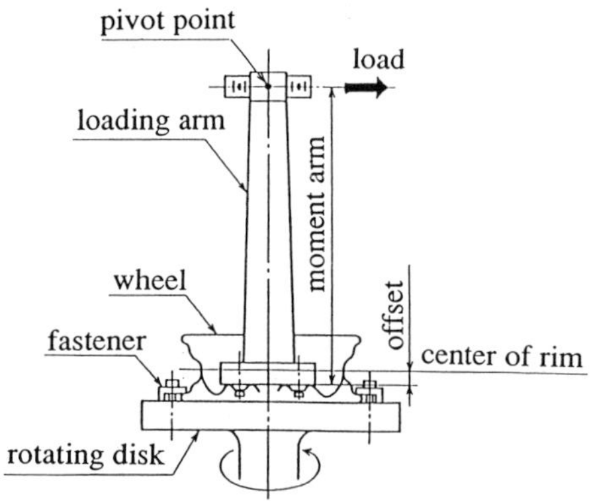

Fig. 3. Rotating bending fatigue test.

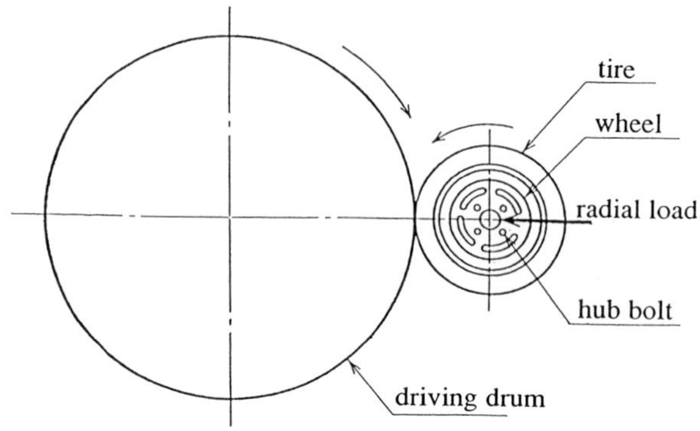

Fig. 4. Radial load fatigue testing. (Drum testing)

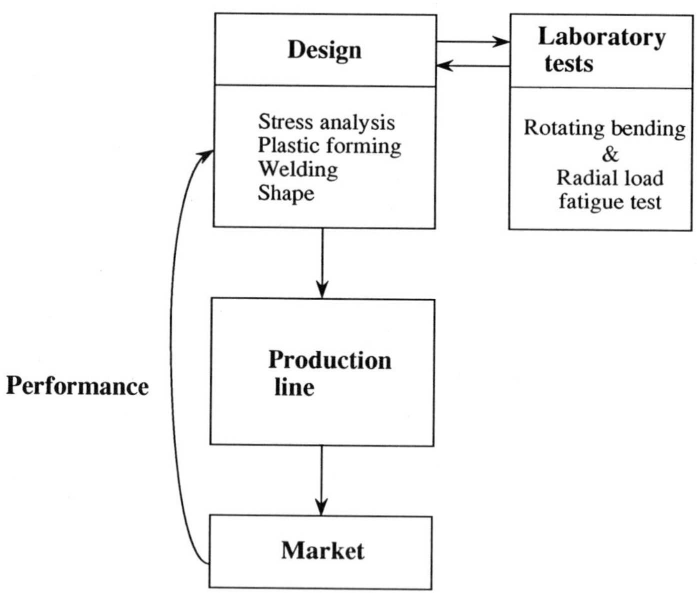

Fig. 5. Conventional fatigue design.

Another type of conventional fatigue design of car wheels is based on the road test in which the radial road on wheels is measured. The data of radial load are applied in laboratory test. However, this method is very complicated and is not enough convenient to conduct the test for every new model.

To avoid the drawback of this conventional design method, the strain measurement of car wheels has been carried out mostly by a system consisting of a combination of slip rings and strain gauges or with a wireless FM telemeter. Then problems with noise induced by use of slip rings and FM telemeter are experienced.

In order to solve these problems, we have established a new fatigue design method. It may be illustrated as Fig. 6. In addition to the conventional system, we put the data acquisition procedure between the laboratory test and market. In the data acquisition procedure, we conduct the strain measurements by road tests in which Mini-Rainflow Corder is attached directly to the central part of a wheel as Fig. 7. The recorder is connected with the lead wire to a strain gauge attached at the critical location. The critical locations can be found from laboratory testing. The cumulative frequencies of strain obtained from road tests are then used to determine, by Miner's Rule, the cumulative fatigue damage D at the critical location for a particular mileage (say 10^5km and 10^6km). This procedure may be repeated until we can assure the fatigue strength reliability of a car wheel model by modifying or revising its shape, plate thickness, welding and surface treatment.

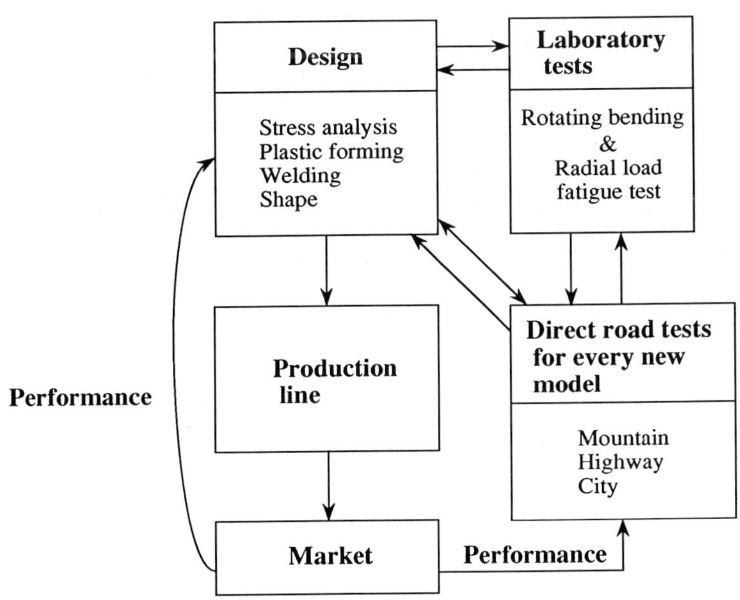

Fig. 6. New fatigue design of car wheels.

Fig. 7. Mini-Rainflow Corder attached to the front wheel.

APPLICATION

An application to the development of a new model of car wheel will be described. Based on laboratory tests and data acquisition by road tests at (a) mountain road, (b) highway and (c) city roads, a prototype was modified three times.

Figure 8 shows the critical location of fatigue crack initiation which was detected by a laboratory test on the 2nd model. The critical locations were the top of the hat and the point of welding between disk and rim. Strain gauges were attached at these critical locations.

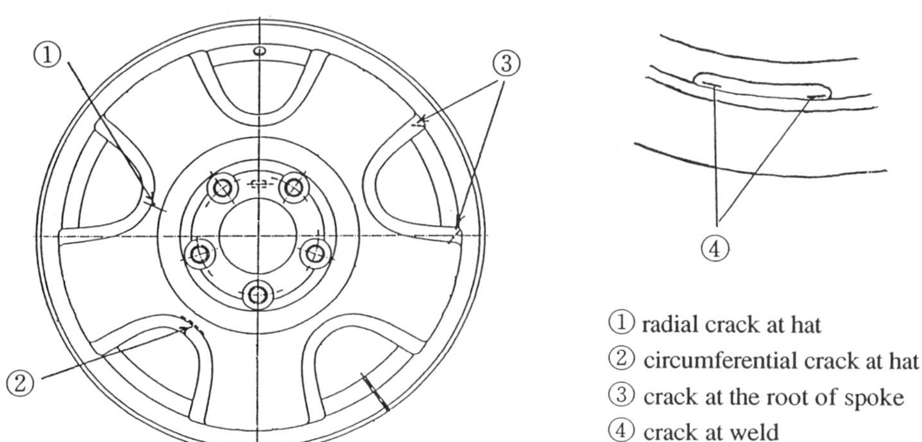

Fig. 8. Rough illustration of typical crack initiation sites.

Figure 9 shows Strain ($\Delta\varepsilon$)–Fatigue life (Nc) curve obtained by the drum test (radial load fatigue test). $P1$ and $P2$ in Fig. 9 mean the loads applied in the drum tests. Nc was defined by the initiation of a 2mm crack. The $\Delta\varepsilon - Nc$ curve was drawn using the average data for 3 tests at each load (P).

Figure 10 shows $\Delta\varepsilon$–Nc curve obtained by the rotating bending fatigue test. $M1$ and $M2$ mean the bending moment applied in the rotating bending fatigue tests. $\Delta\varepsilon$–Nc curve was drawn using the average data points for 9 tests at bending moment $M1$ and 3 tests at $M2$.

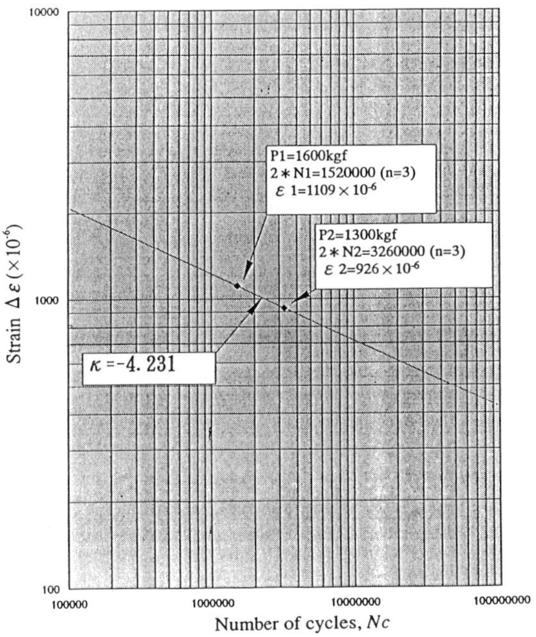

Fig. 9. Radial load fatigue test.
4th model (Final model).

Nc : Cycles
(Reversals : $2\times N1$ and $2\times N2$)
to crack initiation at the welded part.

κ : Slope of $\Delta\varepsilon$–Nc curve.

n = 3 : Numbers of specimens.

Crack initiation cycles :

N1= 5.94×10^5, 7.94×10^5, 8.94×10^5 at P1=1600kgf.

N2= 1.2×10^6, 1.7×10^6, 2.0×10^6 at P2=1300kgf.

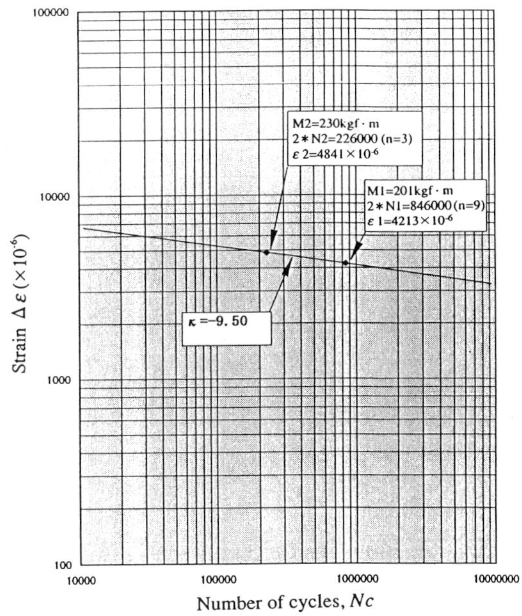

Fig. 10. Rotating bending fatigue test
(Cornering fatigue test).
4th model (Final model).

Nc : Cycles
(Reversals : $2\times N1$ and $2\times N2$)
to crack initiation at the top of hat.

κ : Slope of $\Delta\varepsilon$–Nc curve.

n = 3 : Numbers of specimens.

Crack initiation cycles :

N1= 3.98×10^5, 2.46×10^5, 4.76×10^5, 2.30×10^5, 4.61×10^5, 3.59×10^5, 11.25×10^5, 4.61×10^5, 4.71×10^5 at M1=201kgf · m.

N2= 9.2×10^4, 1.35×10^5, 1.11×10^5 at M2=230kgf · m.

In Figs.9 and 10, there are some scatters in the data of 3 or 9 tests. Therefore, this scatter of the data must be considered in the evaluation of life by the road test. In Figs.9 and 10, fatigue limit is not assumed and the straight *S-N* curves were used by counting fatigue damage based on the modified Miner Rule, because using a monotonically decreasing *S-N* curve is more rational for fatigue under variable amplitude loading than using conventional *S-N* curve with knee point.

Table 1 shows the frequencies of strain measured by the city road test for 14.2km drive of the prototype. The table presents the frequency of strain by every minimum unit of 27×10^{-6} along the horizontal row and by every minimum step of 10 times of 27×10^{-6} along the vertical column.

Table 1. Frequency of strain measured by Mini Rainflow Corder (prototype)
City road, 14.2km

	0	1	2	3	4	5	6	7	8	9
0 ×10	2	85	73	18	12	18	12	14	48	36
1 ×10	48	122	172	348	1303	2449	3299	3378 *	1768	836
2 ×10	338	216	99	69	42	20	18	26	10	8
3 ×10	6	10	2	2	4	4	0	0	4	2
4 ×10	4	0	2	0	0	0	0	0	0	0
5 ×10	0	0	2	0	0	0	0	0	0	0
6 ×10	0	0	0	0	0	0	0	1	0	0
7 ×10	0	0	0	0	0	0	0	0	0	0
8 ×10	0	0	0	0	0	0	0	0	0	0
9 ×10	0	0	0	0	0	0	0	0	0	0
10 ×10	0	0	0	0	0	0	0	0	0	0
11 ×10	0	0	0	0	0	0	0	0	0	0
12 ×10	0	0	0	0	0	0	0	0	0	0
13 ×10	0	0	0	0	0	0	0	0	0	0
14 ×10	0	0	0	0	0	0	0	0	0	0
15 ×10	0	0	0	0	0	0	0	0	0	0
16 ×10	0	0	0	0	0	0	0	0	0	0
17 ×10	0	0	0	0	0	0	0	0	0	0
18 ×10	0	0	0	0	0	0	0	0	0	0

* For example, the number "3378" shows that strain of $(17+0.5)\times27\times10^{-6}$ $(=472.5\times10^{-6})$ was measured 3378 times for 14.2km drive at a city road, where 0.5 of (17+0.5) means that we take the average value between the strain level of 17 and 18.

Figure 11 shows the frequencies (cycles) of strain arranged from Table 1 for a 10km drive.

Figure 12 shows the fatigue damage per 10km drive which was calculated using the $\Delta\varepsilon - N_c$ curve of Fig. 11. The difference between fatigue damage from a mountain road, highway and city road is distinct. These data were used to develop and assure the fatigue strength reliability of the final model which was intended to be put into the market place.

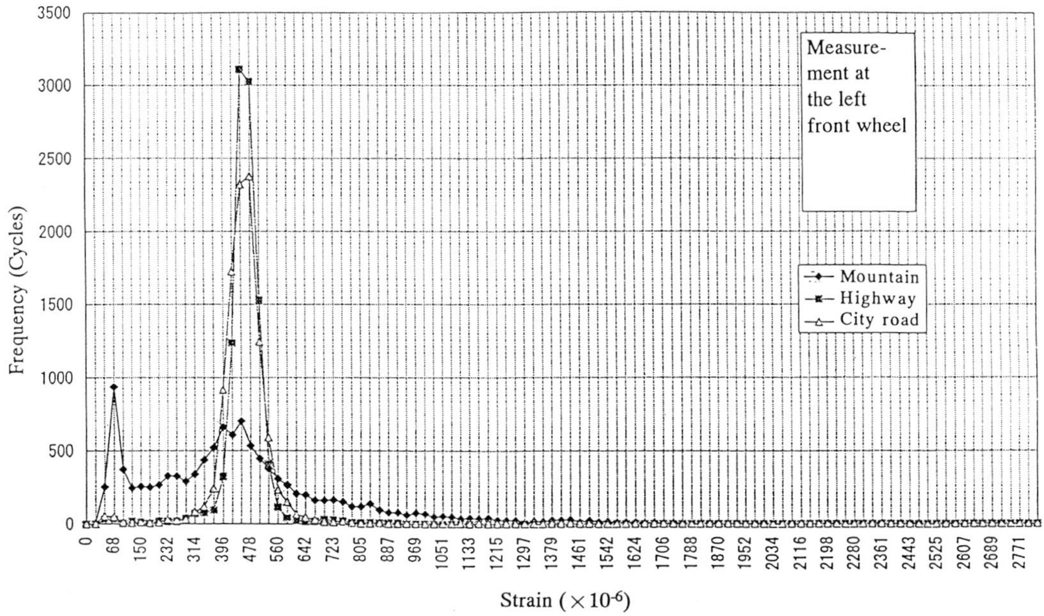

Fig. 11. Comparison of frequency of strain per a 10km drive (prototype).

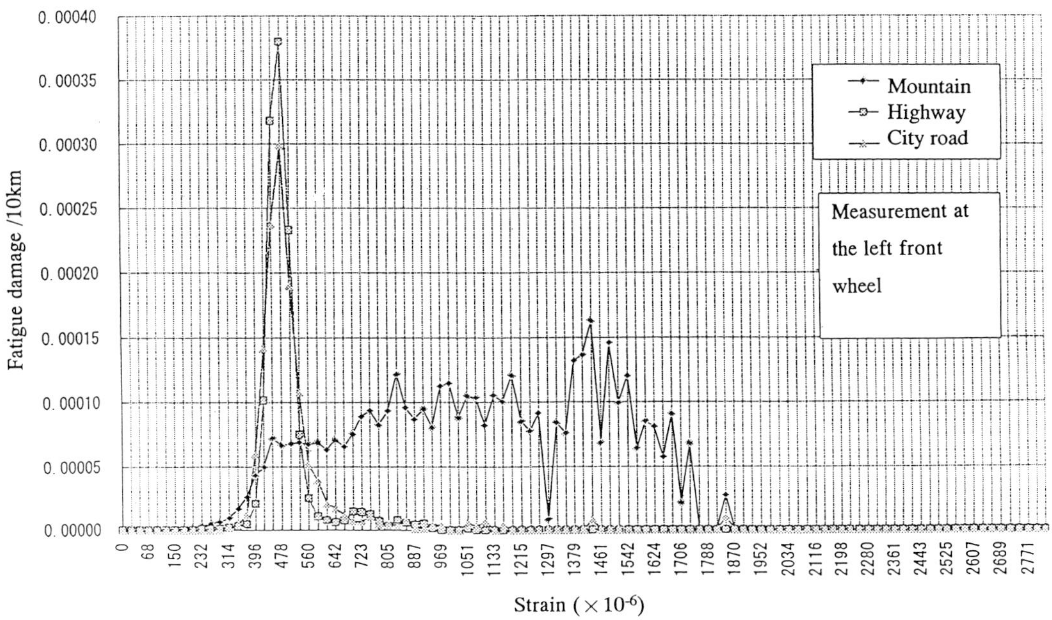

Fig. 12. Comparison of fatigue damage per a 10km drive (prototype).

The flow of the development was as follows.
Table 2 shows the transition of the main specifications from the prototype to 4th model.
Mini-Rainflow Corder was used for the strain measurement at the critical positions of all models during the development process.

Table 2. Transition of the main specifications and modifications from the prototype to the final model

	prototype	2nd	3rd	4th
Plate thickness of disk (mm)	4.5	4.5	4.5	4.5
Plate thickness of rim	3.0	3.5	3.5	3.5
Length of arc weld (mm)	60	80	80	80
Shape design	—	Total change	Partial change (Hole)	Same as 3rd
Shot peening on weld	No	Yes	Yes	Yes
Shot peening on hat	No	No	No	Yes

(1) Prototype
After rotating bending fatigue test and drum testing, road tests were carried out using Mini-Rainflow Corder to investigate the durable mileage in the market. The car used for the data acquisition was 1550kg in total weight (750kg on the front axle and 800kg in the rear). The car was rear wheel right hand drive and contained a single passenger. Strain gauges were attached to all four car wheels. In the data acquisition process, it was found that the strain at the front right hand side wheel had the largest value. The fatigue life at the arc-welded location for mountain road driving was estimated as only $\sim 10^3$km.

(2) Second model
In order to improve the fatigue strength at the arc welded location, the shape of the wheel was modified and the plate thickness of rim was increased. Moreover, shot peening was applied to the weld. The shape change, i.e. reduction of hole size, was made to increase the length of weld. However, this modification caused the crack initiation site to moved to the hat during the laboratory test. The fatigue life in mountain road duty was estimated as $\sim 10^4$km by the road test.

(3) Third model
Additional modifications, the change of hole shape and application of shot peening to weld were then made. The fatigue life at the hat and welded part, by city road testing was then estimated by the measurement by Mini-Rainflow Corder as $\sim 10^5$km. However, the life at mountain road was still estimated as $\sim 10^4$km.

(4) 4th model (Final version)
After additional modifications, harder shot peening to the weld and shot peening to the hat were made, strain measurement on a mountain road was carried out. The life on mountain road was estimated as $\sim 10^6$km and the model was regarded as the final version of the new car wheel.

CONCLUSION

The newly developed small portable strain histogram recorder was applied to the fatigue design of car wheels. The conclusions obtained are as follows.
(1) The new recorder named Mini-Rainflow Corder, is useful for data acquisition. It negates the use of slip rings and FM telemetry.
(2) The strain cumulative frequency obtained during the road test showed a distinct difference in fatigue damage to that obtained during mountain road, highway and city road driving.
(3) The new fatigue design method using Mini-Rainflow corder in combination with the conventional method using rotating bending fatigue test and axial load drum testing has been established. An example of the development process from the prototype to the final version has been described.

ACKNOWLEDGMENT

The present study has been conducted as the joint research project of industries and universities organized and supported by Fukuoka Industry, Science and Technology Foundation.

REFERENCES

1. Murakami, Y., Morita, T. and Mineki, K. (1997). Development and Application of Super-Small Size Strain History Recorder Based on Rainflow Method, *J. Soc. Materials Sci., Japan*, **46-10**, pp. 1217-1221.
2. Matsuishi, M. and Endo, T. (1968). Fatigue of Metals Subjected to Varying Stress-Fatigue Lives Under Random Loading, *Preliminary Proc. of The Kyushu District Meeting, Japan Soc. Mech. Engrs*, pp. 37-40.
3. Endo, T., Matsuishi, M., Mitsunaga, K., Kobayashi, K. and Takahashi, K. (1974). Rainflow Method, the proposal and the applications, *Memoirs of Kyushu Inst. Tech.*, **28**, pp. 33-62.
4. Anzai, H. and Endo, T. (1979). On-site indication of fatigue damage under complex loading, *Int. J. Fatigue,* **1,1**, pp. 49-57.
5. Anzai, H., (1992). Algorithm of the Rainflow Method, *The Rainflow Method in Fatigue, ed. Murakami. Y., Heinemann Butterworth,* pp. 11-20.

ON THE NEW METHOD OF THE LOADING SPECTRA EXTRAPOLATION AND ITS SCATTER PREDICTION

M. NAGODE & M. FAJDIGA
Faculty of Mechanical Engineering, Aškerčeva 6, 1000 Ljubljana, Slovenia

ABSTRACT

Loading spectra are of paramount importance for service life prediction and durability approval. To obtain the necessary data, load measurements or simulations have to be carried out. The influence of variable operating conditions upon loading spectra should be taken into account as well. The present paper deals with the properties of a general multi-modal probability density function (p.d.f.) for load ranges of stationary random processes. Additionally, a new, more general method for prediction of the scatter of loading spectra is presented. However, the main emphasis of the article is the proof that the procedures mentioned above hold also in the case of variable operating conditions. To illustrate the advantages of the new method compared to the existing one, the data obtained by measurements on a fork-lift at various operating conditions have been analysed by both methods independently.

KEYWORDS

Loading spectra extrapolation, variable operating conditions, multi-modal Weibull p.d.f., conditional p.d.f. of load ranges, conditional p.d.f. of the number of load cycles.

INTRODUCTION

Service life is rarely estimated directly from load (stress) time histories. Most often load time histories are first transformed into loading spectra by using counting methods. Then the correlation between the parameters characterizing random behaviour (number, magnitude, sequence of load cycles etc.) and service life is studied [1, 2]. To predict load ranges, the load range distribution function has to be assigned to the corresponding measured or simulated loading spectra.

It has turned out that the currently available distribution functions are not appropriate for all loading spectra. Therefore it has been necessary to find out a distribution function that would be suitable for any loading spectrum. To model loading spectra, a general multi-modal Weibull distribution function [3] has been suggested and tested

$$F(s) = 1 - \sum_{l=1}^{m} w_l \exp\left(-\left(\frac{s}{\theta_l} \right)^{\beta_l} \right). \qquad (1)$$

The function consists of m Weibull distributions. Constant m stands for the number of Weibull distributions, w_l is used as the weighting factor, β_l and θ_l represent Weibull parameters. Compared to the existing p.d.f. models, the proposed distribution has the following advantages:

1. It can be used for any stationary process.
2. The proposed distribution function is appropriate for all possible shapes of loading spectra. Even the ones combined of more than two basic distributions are not excluded.
3. A straightforward procedure of unknown parameter evaluation has been developed and verified, making the new distribution function very easy to implement.
4. The distribution function of load ranges can be predicted on the basis of short load time history samples. Thus both, the costs and time for experiments or simulations as well as numerical analysis, are lowered/shortened considerably.

A reliable life determination could be carried out only if based on the knowledge of the probability of occurrence for the spectrum. Due to technical and economical reasons, the duration of field measurements is limited, which is why the results of the measurements should be properly extrapolated. Based on a general multi-modal p.d.f. (1) and the theory of extreme values, a new procedure of loading spectra extrapolation and scatter prediction has been developed [4]. In this case the scatter of the loading spectrum is completely determined by two conditional p.d.f.'s: the conditional p.d.f. of load ranges

$$f(s \mid n) = \frac{N!}{(N-n)!(n-1)!} F(s)^{N-n} (1-F(s))^{n-1} f(s) \quad ; \quad (n = 1,\ldots,N) \tag{2}$$

and the conditional p.d.f. of the number of load cycles $f(n|s)$. Conditional p.d.f. $f(n|s)$ is a Gaussian distribution with mean value $m(s)$ and standard deviation $\sigma(s)$, given by

$$m(s) = 1 + (N-1)(1-F(s)) \text{ and } \sigma(s) = \sqrt{(N-1)(1-F(s))F(s)} \tag{3}$$

Variable n stands for the number of load cycles and N stands for the size of the loading block.

The major drawback of the currently available method of the scatter of loading spectra prediction [5] is that it can be used only when operating conditions are fixed. However, using the multi-modal Weibull p.d.f. and conditional distributions $f(s|n)$ and $f(n|s)$, it is possible to overcome this problem. The new method can be implemented in the case of variable operating conditions, too, which is mathematically proved in [6].

VERIFICATION OF THE METHOD

To prove its suitability for the design spectrum prediction, the new method will be compared to the method being used mainly for light-weight design in automotive industry. Design spectrum is the one that should be realized under all possible operating conditions within the expected service life. The basic characteristics of the existing method [7, 8, 9, 10] are as follows:

1. The most important criteria for the design of vital components are the attainment of the expected service life with the required reliability and safety.
2. Service life depends primarily on the loading conditions in service. They can be characterized by the representative design spectrum. The main parameters of the spectrum are usage, structural behaviour and operating conditions. One has to know these parameters in advance to make the application of the method possible.
3. For a reliable evaluation of service fatigue life the probability of occurrence of the design spectrum must be defined. As the usage and operating conditions often vary considerably, the scatter in

service loading is relatively high. For vital components it is therefore absolutely necessary to use a "hard" spectrum that has a very low probability of occurrence.

4. To define the design spectrum, service measurements are needed either to determine the data for different operating conditions or even the customer usage as well. In order to establish realistic design spectrum from field measurements, a suitable measuring programme must be used.
5. The values originating from different loading conditions are separated by filtering. Measured load time histories are extrapolated and partial spectra are worked out. The number of cycles for individual spectra N_i and for maximum values N_{ei} is given as values related to the total number of cycles H_0.
6. The design spectrum is assembled from partial loading spectra.

An Application of the Method on a Fork-lift

Let's assume that the loading conditions of a fork-lift are significantly influenced by four operating conditions: driving forwards and backwards and turning right and left at a maximum speed and maximum additional load. To extrapolate partial loading spectra the multi-modal Weibull distribution (1) will be used. The whole analysis will be carried out for two possible uses of a fork-lift, which will show how simply the probabilities of individual operating conditions in time domain p_i can be varied if the new method is used.

Tables 1. and 3. give the input data that are usually used for preliminary design evaluation and durability approval by tests for non-rotating suspension components [9]. The peak rate of a partial spectrum is given by $v_i = N_i^* / t_i^*$, where t_i^* and N_i^* stand for known values. The former represents the return period of the measured load time history sample, while the later stands for the related size of the loading block. The relation between the relative frequencies in time domain p_i and those of the number of load cycles can now be expressed as $p_i' = v_i p_i / v_p$. Variable $v_p = \sum p_i v_i$ is used as the peak rate of a design spectrum. Index i denotes the i-th operating condition. The input data for different relative frequencies p_i are presented in Tables 2. and 4..

Table 1. Basic data for the derivation of the design spectrum; load case: Z_{LI}

Operating condition	i	p_i	N_i^*	v_i	p_i'	N_{ei}	Distribution
Driving forward	DF	0,70	37	3,700	0,466	$H_{0DF} \cdot 10^{-6}$	Multi-modal Weibull
Driving backwards	DB	0,26	83	8,313	0,389	$H_{0DB} \cdot 10^{-6}$	
Left turn	LT	0,02	202	20,200	0,073	50	
Right turn	RT	0,02	202	20,200	0,073	50	

Table 2. Modified data for the derivation of the design spectrum; load case: Z_{LI}

Operating condition	i	p_i	N_i^*	v_i	p_i'	N_{ei}	Distribution
Driving forward	DF	0,40	37	3,700	0,132	$H_{0DF} \cdot 10^{-6}$	Multi-modal Weibull
Driving backwards	DB	0,20	83	8,313	0,148	$H_{0DB} \cdot 10^{-6}$	
Left turn	LT	0,20	202	20,200	0,360	50	
Right turn	RT	0,20	202	20,200	0,360	50	

Figures 1.a. and 1.b. present the assembly of the design spectrum according to both, the present and the new method. Small discrepancies between the spectra are due to: firstly, the imperfection of the

algorithm for calculating the unknown constants of the multi-modal Weibull distribution, and secondly, to the short time history samples ($t_i^* = 10\text{s}$) which the loading spectra have been extracted from (the presence of random fluctuations in the spectra).

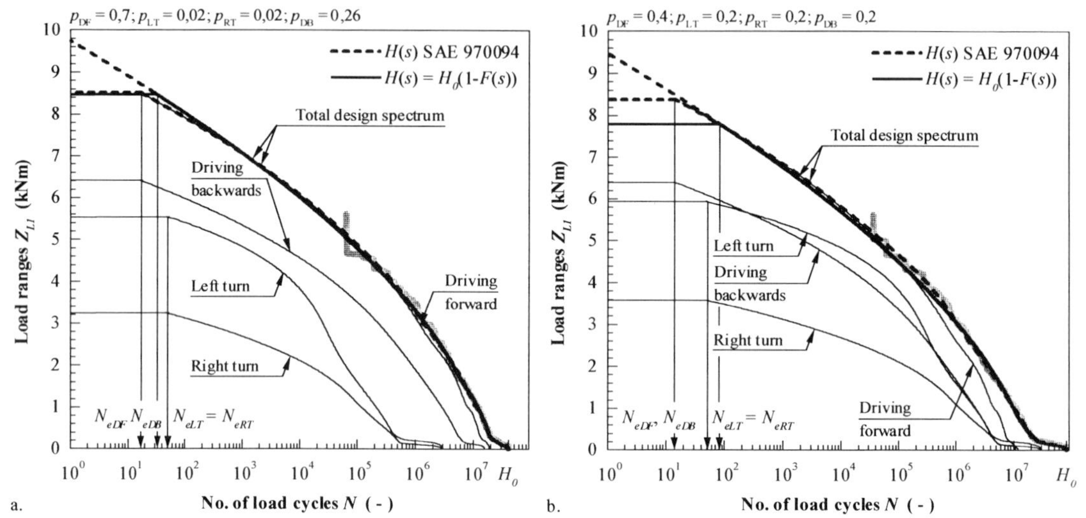

Fig. 1. Design spectrum assembled by using both methods; load case: moment representing vertical force acting on the front left wheel Z_{L1}

Table 3. Basic data for the derivation of the design spectrum; load case: Y_{L1}

Operating condition	i	p_i	N_i^*	v_i	p_i'	N_{ei}	Distribution
Driving forward	DF	0,70	62	6,150	0,493	$H_{0DF} \cdot 10^{-6}$	Multi-modal Weibull
Driving backwards	DB	0,26	139	13,850	0,413	$H_{0DB} \cdot 10^{-6}$	
Left turn	LT	0,02	99	9,867	0,023	50	
Right turn	RT	0,02	311	31,086	0,071	50	

Table 4. Modified data for the derivation of the design spectrum; load case: Y_{L1}

Operating condition	i	p_i	N_i^*	v_i	p_i'	N_{ei}	Distribution
Driving forward	DF	0,40	62	6,150	0,183	$H_{0DF} \cdot 10^{-6}$	Multi-modal Weibull
Driving backwards	DB	0,20	139	13,850	0,206	$H_{0DB} \cdot 10^{-6}$	
Left turn	LT	0,20	99	9,867	0,147	50	
Right turn	RT	0,20	311	31,086	0,463	50	

Larger discrepancies between the spectra appear only because the maximums of the loads are limited. By using the present method, maximum values are defined for each partial spectrum separately. To shorten the procedure for the determination of distribution functions and the extrapolation of partial loading spectra, maximum loads (upper threshold limits) should be defined in a different way. In our case we have decided that the maximum load value will be represented by the value that would be realized $H_0 \times 10^{-6}$ times. The same condition is characteristic of the present method for driving straight on. Constant H_0 stands for the expected service life or the size of the loading block. The comparison of the design spectra for the case of side force Y_{LI} shows agreement as good as in the former case (see Fig. 2.).

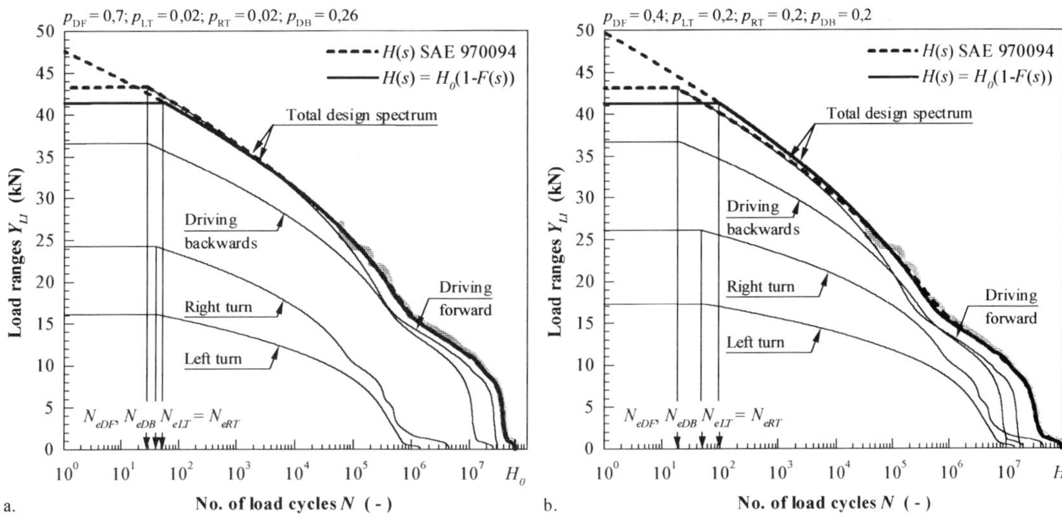

Fig. 2. Design spectrum assembled by both methods; load case: side force acting on the front left wheel Y_{LI}

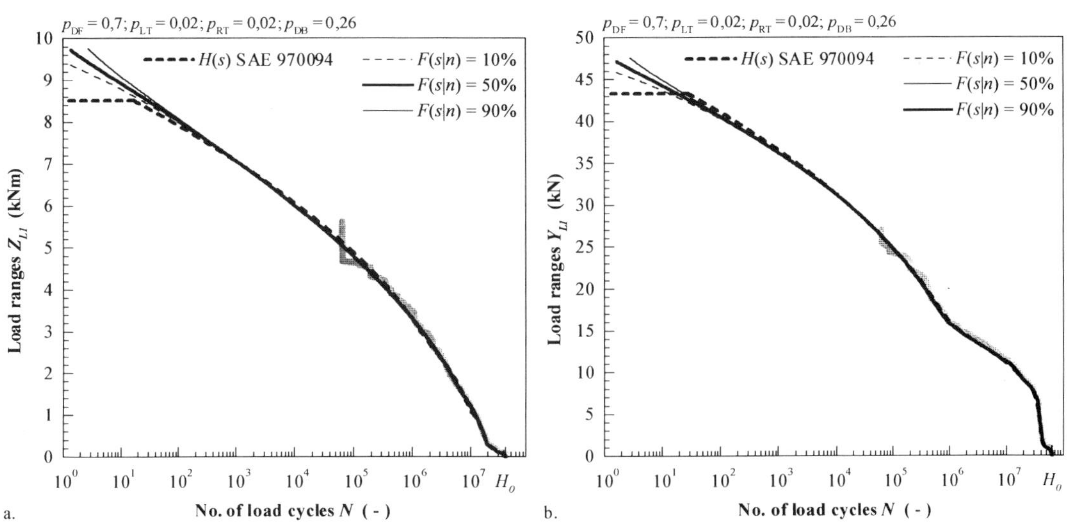

Fig. 3. The scatter of a representative design spectrum prediction; load case: a.) moment representing vertical force acting on the front left wheel Z_{LI}, b.) side force acting on the front left wheel Y_{LI}

Comparing the present method to the proposed one results in the following conclusions:

1. Design spectrum determination is much easier and quicker, for it is only the distribution of

representative design spectrum that has to be defined. The p.d.f.'s of partial spectra may be neglected. Therefore the analysis can take into account a much larger number of different operating conditions instead of including just some of the most inconvenient ones.
2. If field measurements are performed in such a way that the record lengths belonging to particular operating conditions are taken in proper proportions, the filtering of load time histories is even not necessary. From the measured load time history that is characteristic, say, of a representative driver only one representative design spectrum has to be worked out.
3. It has been proved [6] that the design spectrum is also distributed with the same distribution if partial spectra are distributed according to the multi-modal Weibull distribution. The only difference between the two distributions is in the values of unknown constants w_l, β_l and θ_l. Since the p.d.f. of the design spectrum is the same as that of partial spectra, the scatter of the representative design spectrum can be determined by using equation (2). Figure 3. depicts an example of the scatter of a representative design spectrum prediction.
4. By the existing method [5] the scatter of a representative design spectrum can be predicted only when the spectrum shape is linear. This limitation is done away with if the existing method is replaced by the proposed one.
5. However, the new method can not be used in all cases. For instance, when the medium load level changes considerably according to the chosen operating condition, load time histories should be filtered and partial spectra should be treated separately. Similarly, the advantages of the proposed method can not be fully used if the damage mechanism is a combination of low-cycle and high-cycle fatigue.

CONCLUSIONS

In the article only an overview of the proposed method for representative design spectrum prediction has been presented. All the details necessary to bring the method into use have been thoroughly described in [3, 4, 6]. It is rather difficult to treat the influence of a greater number of different operating conditions upon a design spectrum. Working machines generally offer a greater number of variable operating conditions than other vehicles. Therefore it was a fork-lift that has been chosen to be the testing object.

The parameter estimation for the multi-modal Weibull distribution is not as straightforward as the parameter estimation for distributions used presently. The same is true also for the scatter of loading spectra prediction. However, with the proposed method parameters and scatter may be worked out just for a single spectrum. Consequently, the design spectrum prediction is much faster and simpler, which is most probably the major advantage of the proposed method.

REFERENCES

1. Collins, J.A. (1993). *Failure of Materials in Mechanical Design, Analysis, Prediction, Prevention*. 2nd edn. John Wiley & Sons, New York.
2. Haibach, E. (1989). *Betriebsfestigkeit*. VDI Verlag, Düsseldorf.
3. Nagode, M. and Fajdiga, M. (1998). A General Multi-Modal Probability Density Function Suitable for the Rainflow Ranges of Stationary Random Processes. *Int. J. Fatigue*.
4. Nagode, M. and Fajdiga, M. (1998). On a New Method for Prediction of the Scatter of Loading Spectra. *Int. J. Fatigue*.
5. Buxbaum, O. (1967). *Verfahren zur Ermittlung von Bemessungslasten schwingbruchgefährdeter Bauteile aus Extremwerten von Häufigkeitsverteilungen*. Ph.D. Thesis, LBF, Darmstadt.

6. Nagode, M. (1998). *Distribution Function of Loading States Prediction*. Ph.D. Thesis, University of Ljubljana, Faculty of Mechanical Engineering, Ljubljana.
7. Grubišić, V. (1994). Determination of Load Spectra for Design and Testing. *International Journal of Vehicle Design* vol. 15, no. 1/2, pp. 8-26.
8. Grubišić, V. (1997). Fatigue Evaluation of Vehicle Components - State of the Art, Restrictions and Requirements, Keynote Address to Session "Fatigue Research and Application". *SAE International Congress*, Detroit.
9. Grubišić, V. and Fischer, G. (1997). Methodology for Effective Design Evaluation and Durability Approval of Car Suspension Components. *SAE Technical paper series 970094*.
10. Neugebauer, J. and Grubišić, V. (1987). Zum Betriebsfestigkeitsnachweis von Motorradfahrwerks-komponenten. *VDI Berichte no 657* pp. 407-427.

MATERIAL TESTING FOR FATIGUE DESIGN OF HEAVY-DUTY GAS TURBINE BLADING WITH FILM COOLING

YING PAN, BURKHARD BISCHOFF-BEIERMANN AND
THOMAS SCHULENBERG

Siemens AG Power Generation Group (KWU)
Mülheim a. d. Ruhr, Germany

ABSTRACT

Heavy-duty gas turbine blades, which contain film cooling holes, are subjected to a complicated dynamic mechanical loading. Special material tests have been developed to ensure the long-term operational reliability of the blading. This paper presents results of the testing and appropriate finite element (FE-) calculations, showing that a reliable fatigue life assessment can be achieved with the local approach concept.

KEYWORDS

Heavy-duty gas turbine, film cooling hole, low cycle fatigue, fatigue life, FE-calculation.

INTRODUCTION

A new generation of advanced heavy-duty gas turbines, with substantially increased output and efficiency, has been developed by Siemens[1]. One of the important new technologies applied in this gas turbine, that is well proven in aircraft engines, is extensive film cooling which is used in turbine blading to improve the cooling efficiency. The cooling system of the first-stage blading is illustrated in Fig. 1. As shown, a large number of stress raising film cooling openings have been manufactured into the surface of blades and vanes, where cooling air will be discharged. Non-conventional methods, e.g. laser drilling, EDM (electrical discharge machining), ECD (electrochemical drilling), etc. have been used in drilling processes. These processes can cause micro-cracks on the surface of these openings. This special feature of blading must be investigated and considered in fatigue life design to assure component strength for long-term operation of the gas turbines.

The first stage blades and vanes operate at high gas temperatures. There is a pronounced temperature difference between outside and inside of part walls due to an indispensable intensive cooling. This temperature difference causes thermal stresses at different locations. Additionally, centrifugal forces have also be considered in blades.

Fig.1 First stage blades of heavy-duty gas turbine

Therefore the state of stress of the blading is three-dimensional. It changes during start-up and shut-down processes resulting in low cycle fatigue (LCF) which can limit the operating life of the blading.

As a matter of fact, the LCF life of these components is controlled by those film cooling holes that have a narrow pattern, relatively small diameters and a minor angle to the blade surface which, on the other hand, is required to form a thin film of cooling air on the surface. Moreover, it has also to be taken into account that film cooling holes have generally a rough surface and even micro-cracks.

The life prediction procedure for components with stress concentration can be carried out according to the local approach concept [2,3]. However, for a reliable life prediction, it is not sufficient to determine the maximum local stress and strain via elastic-plastic calculations. Inhomogeneous stress states, manufacturing processes and surface quality should also be taken into account.

In case of the considered blading, the distribution of the LCF loading must be determined by three-dimensional FE-calculations. However, an important step to be checked in the life prediction procedure relates to the application of material design data. This has been determined with standard cylindrical smooth samples. Although consideration has been given to holdtime and to scatter in LCF properties of material, it is still necessary to clarify whether the local approach concept gives a reliable life prediction for the blading under these complex stresses and geometries. Therefore, material tests with special samples together with appropriate FE-calculations have been carried out to verify the design procedure.

EXPERIMENTAL DETAILS

Fig. 2 shows the shape of the LCF sample. Holes were drilled by laser using the same tolerance's and operating parameters of the laser drilling machine as in manufacturing. The thickness of the sample was similar to the wall of the blading. The holes had a pattern with hole density, diameter and angel to surface as found in the leading edge, where a „shower head" cooling is applied, giving the highest cooling hole density.

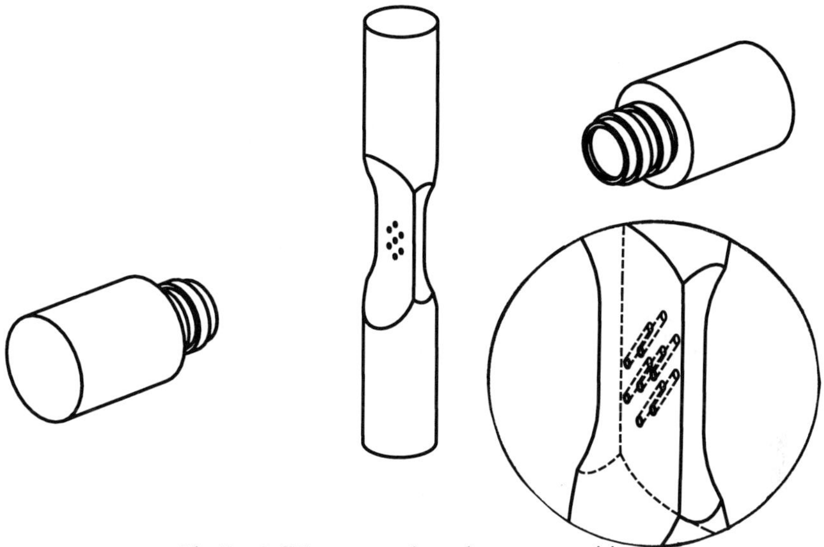

Fig.2 LCF test sample and camera positions

The LCF tests were performed on a servohydraulic test machine with inductive heating under total strain control. The integral strain on the gauge length of the sample was taken as the control signal. The Co-basis superalloy Mar-M509 was investigated at a temperature of 850°C, which is typical for material near the outside on leading edge of first stage vanes during gas turbine operation. The loading of a vane is predominately controlled by thermal strain. This results in compressive strains at the hot outer surface. Therefore the maximum strain in the loading cycles of the LCF test was zero and the minimum strain was varied in different tests. This results in strain ratio $\varepsilon_{min}/\varepsilon_{max}$ equal to $-\infty$.

A camera system was used to continuously observe the surface of samples in the fatigue tests (Fig.2). It was composed of two CCD-cameras with specially developed control programmes so that the crack initiation and propagation could be recorded automatically by computer. After the tests, an evaluation was made for crack behaviour relating to the hole pattern and to micro-cracks. The failure cycles N_f were defined for that number of cycles after which a drop of the stress amplitude of more than 5% was found.

RESULTS AND DISCUSSIONS

Fatigue Test Results

Fig.3 shows the LCF life derived from samples with and without holes. The influence of a holdtime of 20 minutes at maximum pressure loading was also investigated. The smooth samples have the same form shown in Fig.2 but without holes. As no edge influence on crack initiation at the chosen test conditions was observed, these samples gave a very similar number of cycles to failure like that mean value from the standard samples. It can be seen by a comparison based on the integral strain loading that the number of cycles to failure for the samples with holes are lower at all strain levels. A further reduction of the fatigue life was caused by the holdtime in tests, especially at low loading levels, where the experiment duration was very long.

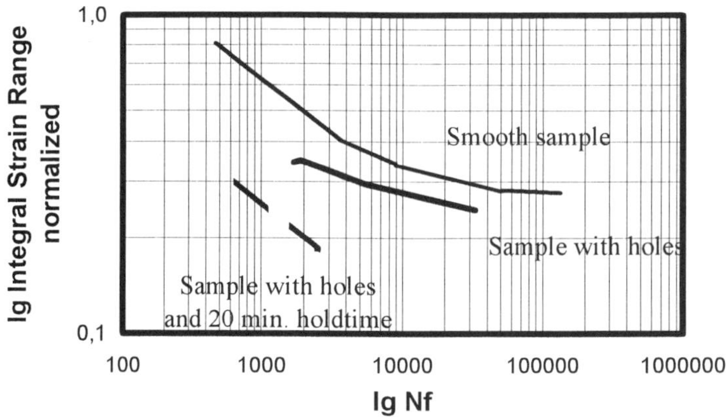

Fig.3 Influence of cooling holes and holdtime on fatigue life

Crack Behaviour

Metallographical investigations were carried out on new samples and samples after failure to analyse the crack behaviour. Micro-cracks beginning from the remelt layer on the surface of laser-drilled holes can be seen in Fig.4(a). The depth is within the tolerance for the normal manufacture procedure. A random distribution of micro-cracks on the hole surface can be seen in Fig.4 (b).

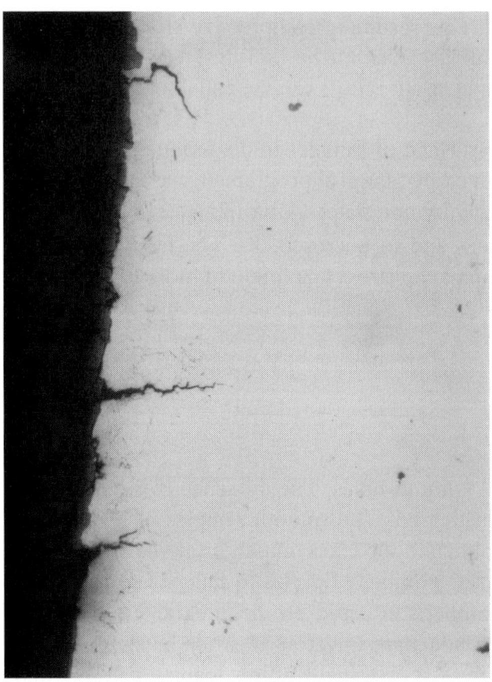

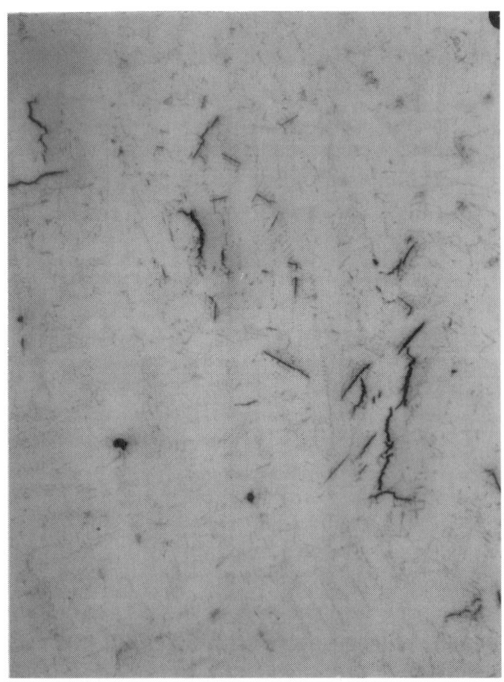

a) Micro-cracks in remelt layer (b) Distribution of Micro-cracks

Fig. 4 Micro-cracks in a new sample after Laser drilling

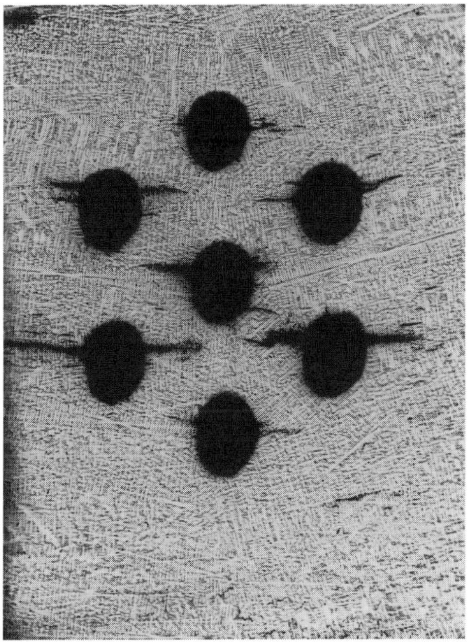

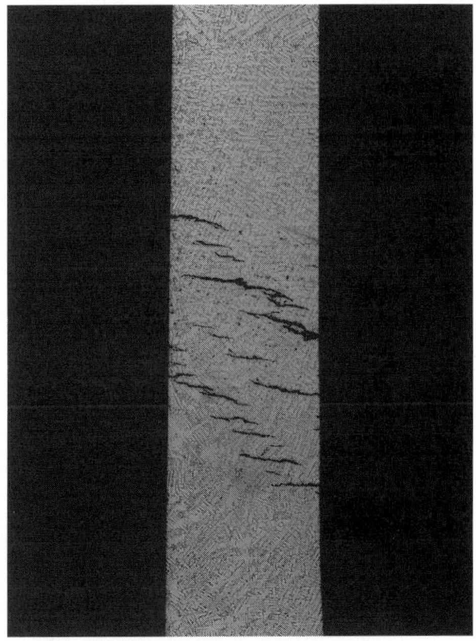

a) Cracks initiated from cooling holes (b) Section between the holes

Fig. 5 Cracks after the complete loading cycles

A crack, initiated by cyclic loading, with a length of more than 0.1 mm on the sample surface could be recorded by the camera system in the fatigue tests. The crack propagation remained slow while more and more cracks arose. With a crack length of about 1 mm the crack propagation slowed further down. Obviously, there was a crack propagation phase of about 95% of the fatigue life for all samples with holes. In this phase, cracks grew from about 0.1 mm to the failure length which corresponded to a stress amplitude drop of 5%. The samples were further loaded as usual until the test machine was automatically shut down by its control system. Fig. 5 illustrates a sample after the final loading cycle which was more than twice the fatigue life.

The sample surface with cracks initiated from the edge of the holes is shown in Fig. 5 (a). After the final loading cycle there were still no direct links of cracks between the holes. The sample was sectioned parallel to the loading direction as shown in Fig.5 (b). The observed cracks were initiated from the inner surface of holes. They grew all perpendicular to the loading direction and formed a crack band parallel to the angle under which the holes were drilled. It is interesting to see that even after much more loading cycles than N_f there are almost no cracks that grew through the whole section.

Finite Element Calculation

The stress concentration at the holes has been calculated with finite elements. Only the part of the test specimen between the extensometer has been modeled. Constant displacements have been applied to the corresponding cross sections between the extensometer to simulate the strain controlled LCF test. The displacement is directly proportional to the nominal strain over the gauge length. The specimen is horizontally symmetrical with an additional axis of symmetry (Fig.6). Therefore, only one quarter of the specimen was modeled. The FE model and the Von-Mises stress, calculated under the assumption of linear elastic material behavior, are shown in Fig. 6 . Highest stresses are found along the curve where the holes intersect the outer surface of the specimen. There is no difference in the distribution and the stress level between the different holes Fig.7. With the help of the FE model, the maximum local stress is calculated as a function of the nominal strain. Based on these results the maximum local total strain (with elastic and inelastic parts) is estimated with Neuber's rule [4-6]. The cyclic stress-strain curve is used to take plastic deformations into account (Fig. 8). From this the relation between the nominal strain over the gauge length and the local strain is derived.

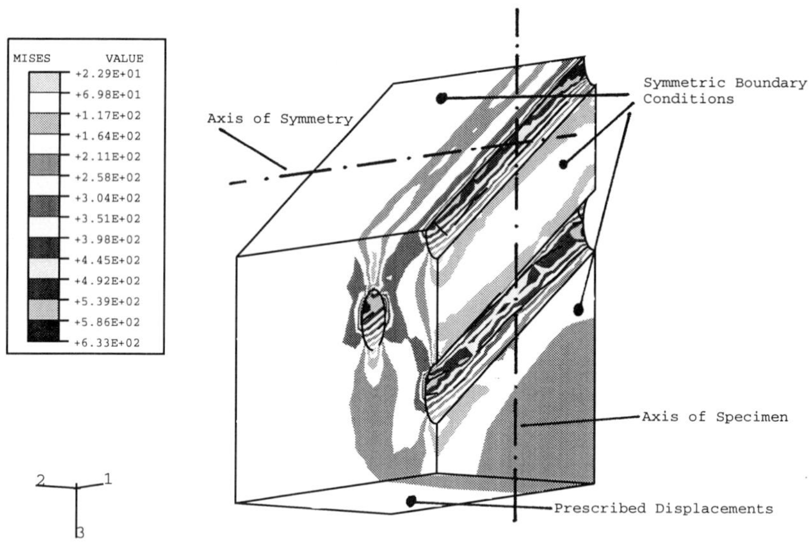

Fig. 6 Stress distribution

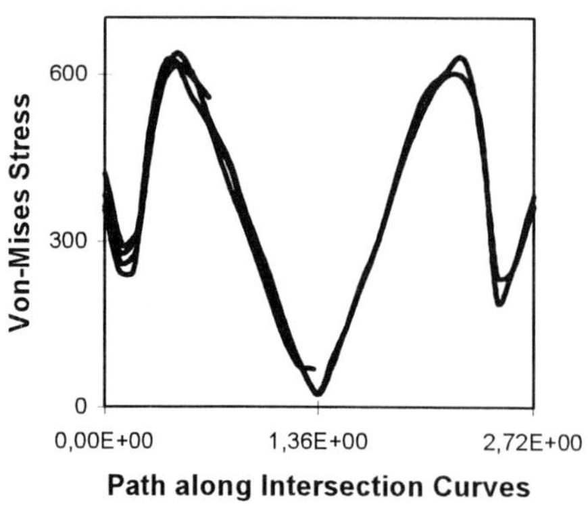

Fig. 7 Distribution of local Stresses along intersection of holes with outer surface

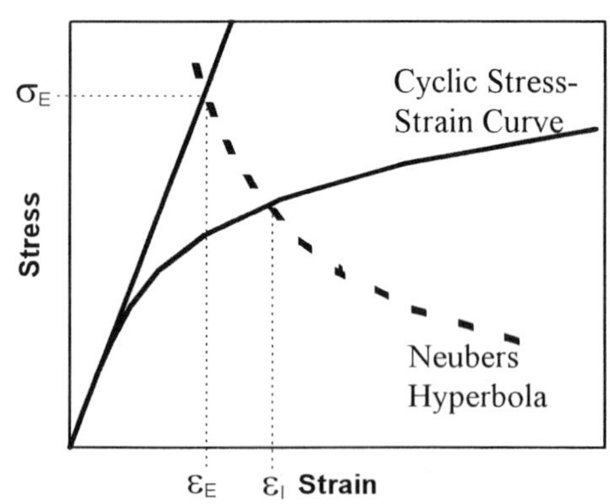

Fig. 8 Calculation of local elastic and inelastic strain

Discussion

S-N curves for samples with holes, which is based on the local strain predicted by the FE-calculations, are presented in Fig.8 and compared with the design curve. The design curve under-estimates the cycles to failure so that a conservative, reliable life prediction using the local approach concept can be expected for the loading conditions considered here. The oxidation effect will be more dominant for the tests with holdtime at lower loading level. Further tests with coatings will be added to represent operation conditions.

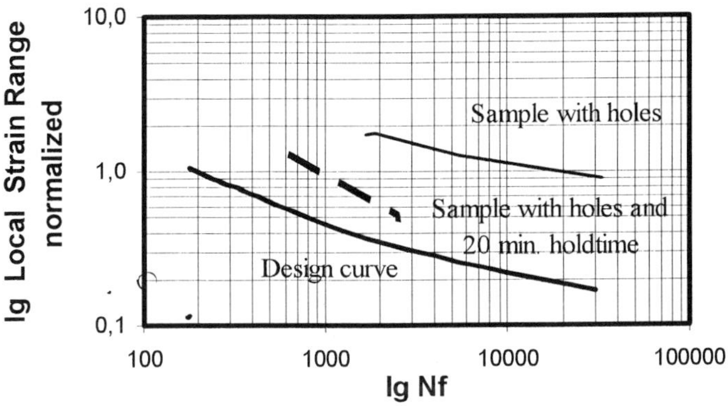

Fig.8 Comparison of results from cooling hole samples with design curve

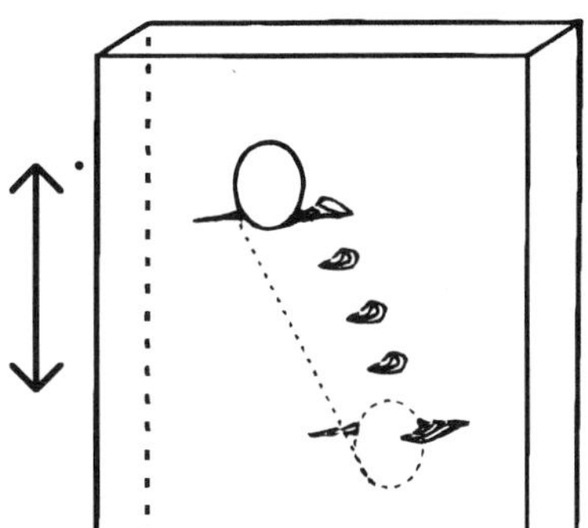

Fig. 9 Crack positions after failure

A simplified depiction in Fig. 9 describes schematically the crack positions in a sample after failure. The cracks begin from the surface of holes and propagate perpendicularly to the main loading direction. The randomly distributed micro-cracks due to Laser drilling have obviously less influence on the crack propagation. The observed reduction of the crack propagation rate during the fatigue tests can be explained by the decrease of stress concentration due to initiation and propagation of many parallel cracks. The crack propagation phase starts after about 5% of the total fatigue life. Before N_f has been reached there is no unstable crack growth to be anticipated. The micro-cracks within the manufacturing tolerance have therefore only a minor influence (less than 5%) on the fatigue life under the considered loading conditions.

The definition of the fatigue life at a 5% drop of the stable stress amplitude is equivalent for standard, strain-controlled tests and for the tests using samples with cooling holes. However, in the latter ones cracks can arise between holes long before the failure.

CONCLUSIONS

- If verified by special LCF fatigue tests using specimens with film cooling holes at appropriate loading conditions, the local approach concept can be used for fatigue design of gas turbine bladings with film cooling holes.
- The local strain can be determined by 3D-FE-calculations with suitable assumptions.
- Micro-cracks formed during Laser-drilling and within the manufacturing tolerance have a minor influence on the fatigue life under the considered loading conditions.
- Crack growth before failure will be slow due to the decrease of the stress concentration in the vicinity of holes. An unstable crack growth is not expected until failure.

REFERENCES

1. B. Becker, T. Schulenberg and H. Termuehlen, The 3A Series Gas Turbines with HBR Combustors, ASME 95-GT-456, Turbo Expo. Houston, Texas, 1995
2. T. Seeger and P. Heuler, Generalised Application of Neuber`s rule. J. Test. Eval., 3, 199-204, 1980
3. G. Savaidis, M. Dankert and T. Seeger, An Analytical Procedure for Predicting Opening Loads of Cracks at Notches, Fatigue Fract. Engng Mater. Struct. Vol. 18 No. 4. 425-442, 1995
4. H. Neuber, Theroy of Stress Concentration for Shear-Strained Prismatical Bodies with Arbitrary Nonlinear Stress-Strain Law, Journal of Applied Mechanics, 12, S.544-550, 1961
5. M. Chaudonneret and J.L. Chaboche, Fatigue Life Prediction of Notched Specimens, Int. Conf. of Fatigue of Engineering Materials and Structures, Sheffield, 1988
6. J. Lemaitre, A Course on Damage Mechanics, Springer Verlag, 1992

CONSIDERATION OF CRACK PROPAGATION BEHAVIOUR IN THE DESIGN OF CYCLIC LOADED STRUCTURES

W. FRICKE, A. MÜLLER-SCHMERL
Germanischer Lloyd, D - 20459 Hamburg, Germany

ABSTRACT

The present practical design of ship structures regarding fatigue is usually based on allowable stress ranges or fatigue lives. These ensure that the probability of crack initiation is sufficiently small. Crack propagation behaviour is not considered. For different kinds of notches the crack propagation rate differs extremely because of different stress gradients along the crack front. By analysis of the crack propagation behaviour for different typical notches, the probability of crack initiation during an inspection interval and crack propagation to a critical state are investigated. The uncertainty of influence parameters and probability of detection of short cracks are considered. The investigation results in a more stringent assessment of notches with large radii. In this way, the consequence of crack initiation and redundancy of the surrounding structure can be rationally taken into account during design.

KEYWORDS

Fatigue, Crack propagation, Notch radius, Fatigue reliability

INTRODUCTION

In recent time ship structures have been designed considering fatigue strength criteria to an increasing extent. In the same way as for steel structures and vehicles, local stress ranges are permitted only up to a certain limit, for which the probability of crack initiation is sufficiently small and, therefore, tolerable. The crack propagation behaviour after initiation - although the latter is assumed to occur rarely - is not explicitly taken into account. Insofar, notches with different radii are regarded to have the same fatigue life as long as they are stressed up to the allowable limit. On the other hand, the crack propagation can be quite different for different kinds of notches because the stress gradient perpendicular to the notch varies. While at relatively sharp notches a crack will soon reach zones with reduced stress concentration and the propagation rate will, therefore, be limited, a crack at mild notches will run through a larger area with high stresses and will reach the critical length earlier (Fig. 1). In ship structures this situation occurs at rounded plate edges with large radii, such as at hatch corners or lightening holes.

In this paper, the probability of crack initiation and reaching a critical length within an inspection interval is investigated by analysing the propagation behaviour of cracks at typical cut-outs with small and large radii. Taking into account uncertain influence factors as well as the probability of detection (POD) for short cracks, conclusions are drawn with respect to a more stringent assessment of rounded plate edges with large radii. In this way, the consequence of a fatigue failure and the redundancy offered by the surrounding material can be considered in fatigue analyses in a simple and rational way.

In a first step the fatigue life of notched members between crack initiation and failure (i. e. reaching the critical crack length) is considered by assuming deterministic parameters and, in a second step, by considering the stochastic parameters as well as inspections.

The objective is to define permissible stress ranges for different notch radii under consideration of the crack propagation behaviour such that the same probability of failure is achieved. The investigation concentrates on differences in crack propagation behaviour only, although the probability of crack initiation of notches with different radii may additionally be affected by size effects.

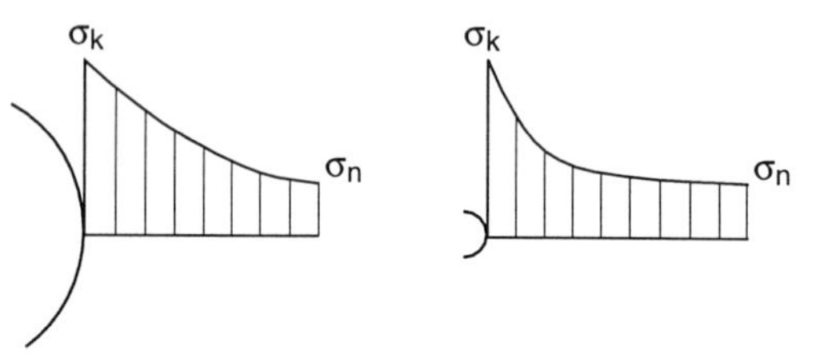

Fig. 1: Stress gradients at different notch radii

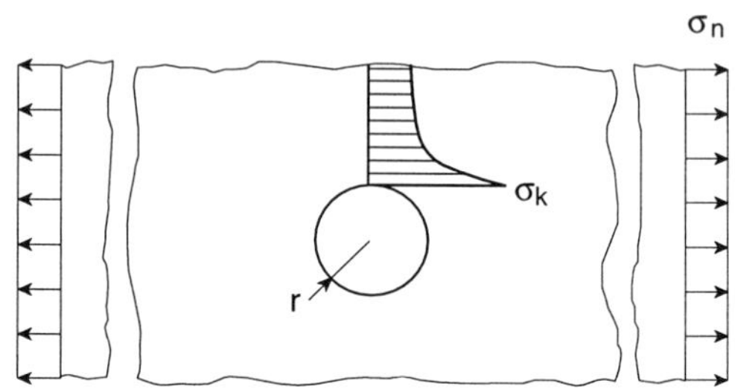

Fig. 2: Investigated plate with a hole

DETERMINISTIC FATIGUE ANALYSIS FOR DIFFERENT NOTCH RADII

In order to investigate the effect of stress distribution at notches on fatigue lives, plates with circular holes having radii r = 50 mm, 150 mm and 500 mm are compared with each other. Infinite plate strips are assumed being subjected to a nominal stress σ_n (Fig. 2). In this case the theoretical notch stress σ_k is the same for all radii. The stress decay beside the hole, which is different for each radius, is regarded as typical for a great variety of rounded notches.

Crack Initiation Phase

When designing the structural detail mentioned according to the Construction Rules of Germanischer Lloyd [1] considering wave-induced stress ranges by the Weibull distribution with a shape parameter of unity (straight line spectrum) and $5 \cdot 10^7$ load cycles (approx. 20 years service life), the largest notch stress range $\Delta\sigma_{k,\max}$ may not exceed the permissible stress range (fatigue resistance) $\Delta\sigma_p$:

$$\Delta\sigma_{k,\max} = K_t \cdot \Delta\sigma_n \leq \Delta\sigma_p \tag{1}$$

The stress concentration factor for circular holes is for all radii $K_t = 3$. According to [1], $\Delta\sigma_p$ results from:

$$\Delta\sigma_p = f_n \cdot \Delta\sigma_R = 526\ \mathrm{N/mm^2} \tag{2}$$

where $\Delta\sigma_R$ = 140 N/mm^2
f_n = 3.76

$\Delta\sigma_R$ is the fatigue strength reference value (detail category) for free plate edges with machine-cut quality, defined for constant amplitude loading at $2 \cdot 10^6$ load cycles and a probability of crack initiation $P_A = 2.5$ %. f_n considers the shape and number of load cycles of the spectrum. Regarding mean stress effects, pulsating tensile stresses are assumed (R = 0).

Crack Propagation Phase

In practical computations, a crack initiation is assumed for a structural member if the crack is visually detectable. The crack length of a so-called "technical" crack is set at 3 mm and corresponds to the smallest crack assumed to be detectable during an inspection of a ship.

The crack propagation is computed using the equation according to Paris and Erdogan:

$$da/dN = C \cdot (\Delta K)^m \tag{3}$$

While C and m are material constants, the stress intensity factor K primarily depends on the crack length a, the geometry of the structure and the applied stress:

$$K = \sigma_n \cdot \sqrt{\pi \cdot a} \cdot Y \tag{4}$$

Schwalbe [2] describes the geometry function Y for a crack at a circular hole by the curve shown in Fig. 3 which has been approximated by the formulae given there.

Taking into account the spectrum due to wave-induced stress cycles by the so-called straight-line spectrum, represented by load sequences with 8 steps (Table 1), the crack propagation behaviour can be analysed by eq. (3) and (4) for different notch radii.

Fig. 4 shows a comparison of computed crack lengths versus the number of load cycles for three different circular holes. The crack initiation phase was assumed to be $5 \cdot 10^7$ load cycles. The differences in propagation behaviour are significant.

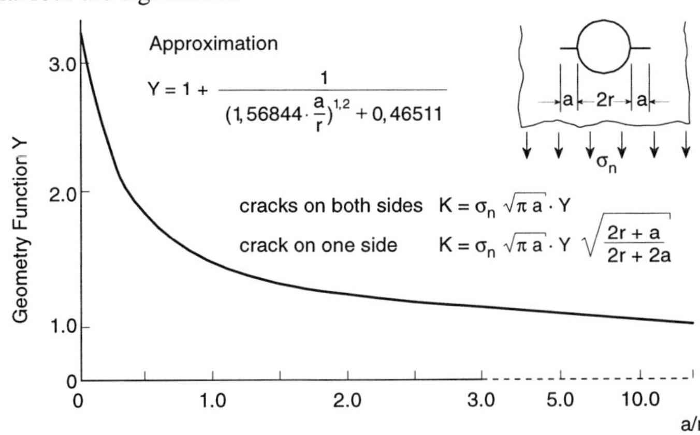

Fig. 3: Stress intensity for cracks at a circular hole in an infinite plate, acc. to [2]

Table 1: Load sequence for straight-line spectrum

Step i	$\Delta\sigma_{n,i}/\Delta\sigma_{n,max}$	No. of load cycles
4	0.625	87
3	0.750	15
2	0.875	3
1	1.000	1
2	0.875	3
3	0.750	15
4	0.625	87
5	0.500	487
6	0.375	2730
7	0.250	15400
8	0.094	462000
7	0.250	15400
6	0.375	2730
5	0.500	487

The critical crack length, from which the crack becomes unstable, is frequently determined by the critical stress intensity factor K_c, so that the following failure criterion applies:

$$K \geq K_c \tag{5}$$

The quantity of K_c depends on material properties, temperature etc. and contains, therefore, high uncertainties. For this investigation, the critical stress intensity factor is assumed to be $K_c = 4000$ N/mm$^{3/2}$. It should be noted that the crack length a and the stress intensity factor K increase rapidly within a few cycles shortly before the critical state is reached, so that the absolute value of K_c only affects the results negligibly.

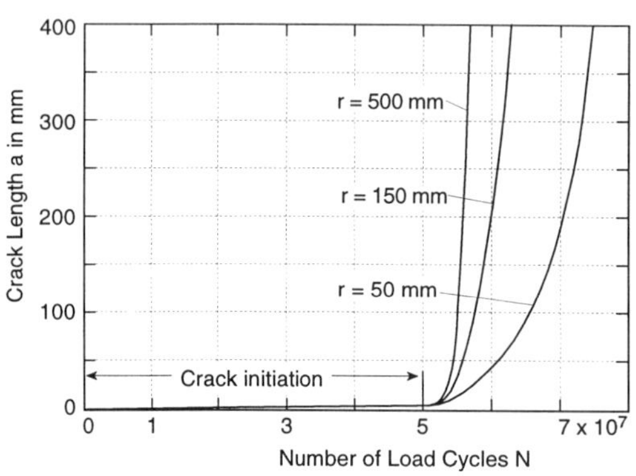

Fig. 4: Crack lengths vs. load cycle number for circular holes

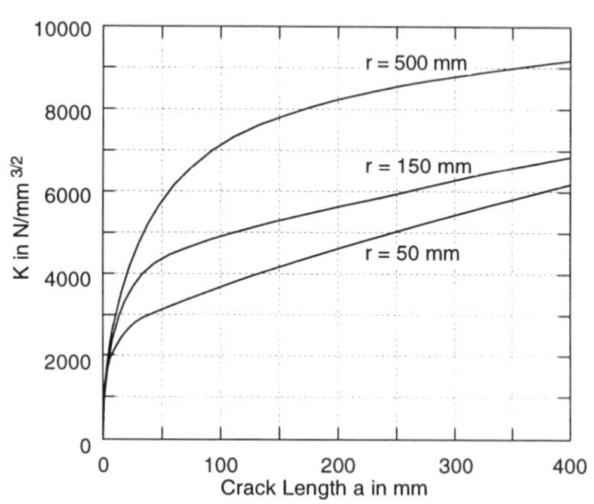

Fig. 5: Stress intensity factor vs. crack length

Fig. 5 shows the relationship between the crack intensity factor and crack length for three notch radii. It becomes clear that the critical state is reached quickest in the structure with the largest notch radius. The deterministic fatigue analysis for the different notch radii yields the critical crack lengths a_c and associated cycles to failure N_C as shown in Table 2.

Table 2: Computed critical crack lengths a_c and number of cycles N_c between crack initiation and failure

r	50 mm	150 mm	500 mm
a_c	138.0 mm	36.3 mm	23.1 mm
N_C	$1.748 \cdot 10^7$	$0.497 \cdot 10^7$	$0.349 \cdot 10^7$

The comparison of the figures shows that the number of cycles in the crack propagation phase, i. e. those between crack initiation and failure, may differ by a factor of five.

Complex Structures

The results obtained for an infinite plate with a hole should also be applicable to more complex ship structures. For lightening holes in plates, this can be justified without further verification. For more complex structures, a hatch corner with r = 500 mm (Fig. 6) was taken as an example and investigated with respect to the crack propagation behaviour.

Loads typical for ship structures were applied with a combination of normal and shear as well as bending stresses in the longitudinal and transverse deck strips as used in [3]. It was assumed that the crack will start from the point with the highest edge stress and propagate perpendicular to the largest principal stress, i. e. perpendicular to the plate edge. Using a very fine finite element mesh, the stress intensity factor K at the crack tip was computed for crack lengths a = 3 mm, 10 mm, 28 mm and 95 mm, applying 8-noded membrane elements and the quarter-point-technique for simulating the stress singularity at the crack tip.

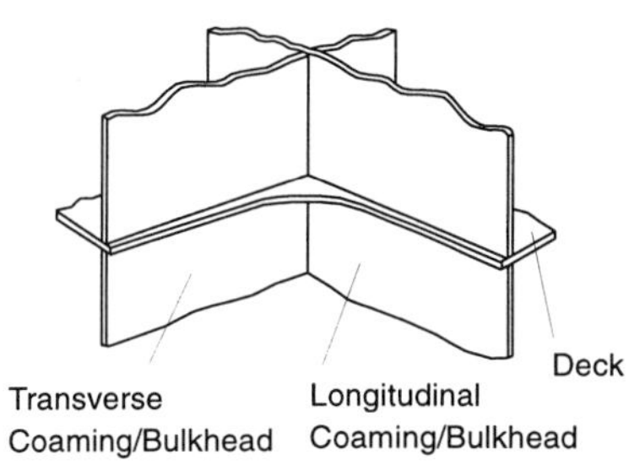

Fig. 6: Hatch corner

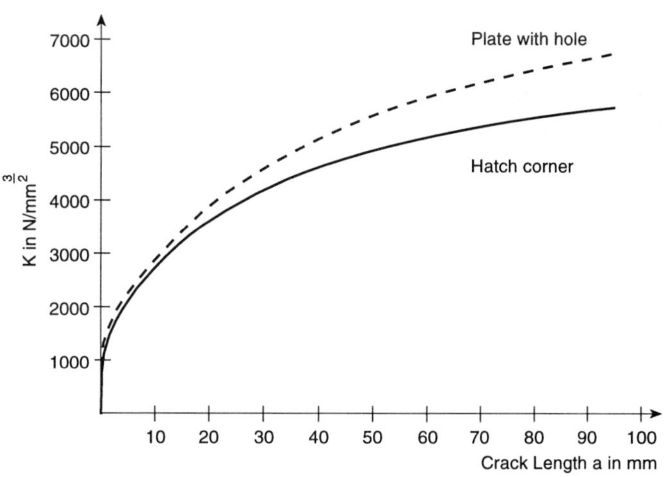

Fig. 7: Stress intensity factor K for a circular hole and for a hatch corner with r = 500 mm

Fig. 7 compares the computed values with those acc. to eq. (4). For the hatch corner, smaller stress intensities are obtained. A reason for this is seen in the restraining effect due to the surrounding structural components (coaming and bulkhead plating). This effect is more pronounced if the crack tip comes close to these components.

In summary it can be stated that the formula given in Fig. 3 is suited for describing the geometry function at rounded corners. For large radii in complex structures like hatch corners, a modified geometry function can be used:

$$Y = \left(0.78 + \frac{0.78}{\left(1.56844 \cdot \frac{a}{r}\right)^{1.2} + 0.46511} \right) \cdot \left(\frac{a}{r}\right)^{-0.04} \tag{6}$$

PROBABILISTIC FATIGUE ANALYSIS FOR DIFFERENT NOTCH RADII

The probabilistic fatigue analysis considers the fact that several variables are affected by high uncertainties and scatter. The statistical parameters of these variables which are used in the analysis are summarised in Table 3. The probability of failure was computed using the Monte Carlo method by simulating a crack initiation for each notch radius 25,000 times.

Table 3: Statistical parameters for the probabilistic crack propagation analysis

Variable	Distribution	Mean Value	Standard Deviation	Source
Crack propagation parameter ln C	normal	-29.84	0.55	[6]
Crack propagation parameter m	constant	3	-	-
Inspection interval (load cycles)	constant	$0.625 \cdot 10^7$	-	-
Number of load cycles per year	constant	$0.250 \cdot 10^7$	-	-
Service life (load cycles)	constant	$5 \cdot 10^7$	-	-
Critical stress intensity factor K_C	normal	4000 $N/mm^{3/2}$	400 $N/mm^{3/2}$	-
Fatigue strength lg $\Delta\sigma_R$	normal	2.146 (140 N/mm^2)	0.0630	[1], [4]
Max. local stress range lg $\Delta\sigma_{k,\,max}$	normal	2.721 (525.7 N/mm^2)	0.041	-

The probability P_f for the failure of a structure results from the probability P_A for a crack initiating during the service life and the probability P_C for reaching the critical state during propagation:

$$P_f = P_A \cdot P_C \tag{7}$$

Probability of Crack Initiation

As described in the deterministic fatigue analysis, the same number of endurable stress cycles and, therefore, the same probability for crack initiation is assumed for all different notch radii. This is in agreement with common rules and guidelines for fatigue strength assessment assuming the same reliability for all detail categories, although the probability for crack initiation at a large radius may be higher than at a smaller radius, as mentioned in the introduction.

During the service life, the fatigue strength steadily decreases and the probability of crack initiation increases as demonstrated schematically by the life curve in Fig. 8. The scatter band of the fatigue strength has been assumed to be described by the following ratio between 90 % and 10 % survival probability:

$$T_\sigma = \Delta\sigma_{90\,\%} / \Delta\sigma_{10\,\%} = 1 : 1.45 \tag{8}$$

Haibach [4] proposes a ratio of 1 : 1.26 for the fatigue strength of base metal. The scatter of the life curve given in eq. (8) is increased because it contains the additional scatter and uncertainties related to variable amplitude loading. On the load side, the variation coefficient of the stress ranges was assumed to be 10 %.

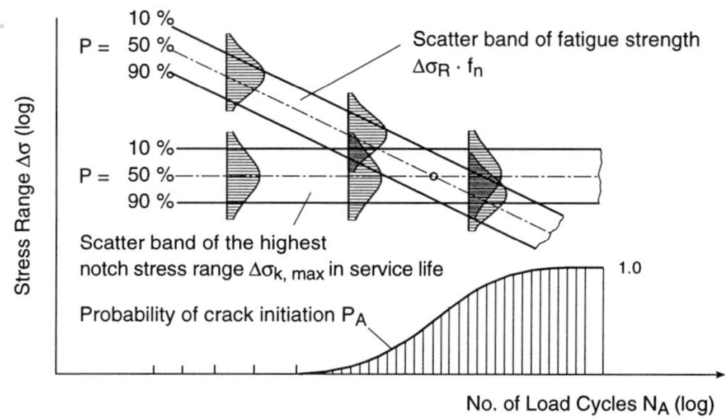

Fig. 8: Schematic representation of the probability of crack initiation due to the scatter of applied stresses and fatigue strength

Using these values, the cumulative probability of crack initiation during the lifetime shown in Fig. 8 was computed with the program COMREL [5] using the assumptions given in the construction rules, i. e. a total probability of crack initiation $P_A = 2.5$ % after approx. 20 years service life. The vertical lines in Fig. 9 indicate the inspection intervals. For structural components in ships, an inspection interval of 2.5 years (approx. 6.25 million load cycles) is regarded as typical.

For the analysis performed here, the curve is linearized within the individual inspection intervals, i. e. assuming a constant rate of crack initiation within the intervals.

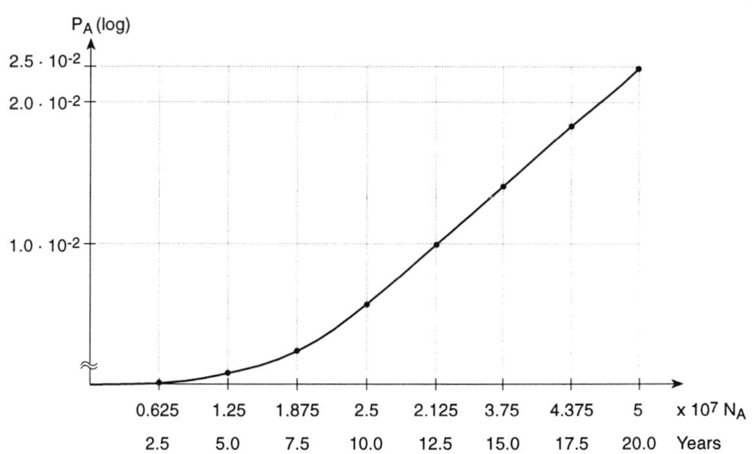

Fig. 9: Computed probability of crack initiation vs. service life

Computation of the Failure Probability

Based on these rates 25,000 crack initiations were simulated over the service life of approx. 20 years. For the subsequent crack propagation until failure, mean cycle numbers between $1.67 \cdot 10^7$ and $0.34 \cdot 10^7$ were computed for the notch radii 50 mm, 150 mm and 500 mm (see also Table 4), using the statistical data given in Table 3. Similar results have been obtained in the deterministic analysis (Table 2).

The possibility that the crack may be detected during an inspection before it reaches the critical state is important in this connection. In the case of slow crack propagation being typical for relatively sharp notches, for instance, it may be expected that cracks initiating during an inspection interval are detected with high probability at the next inspection before reaching the critical length. The inspection scheme used for the computations is shown in Fig. 10. It assumes that every crack detected will be repaired. On the other hand, it may happen that existing cracks are not detected during an inspection.

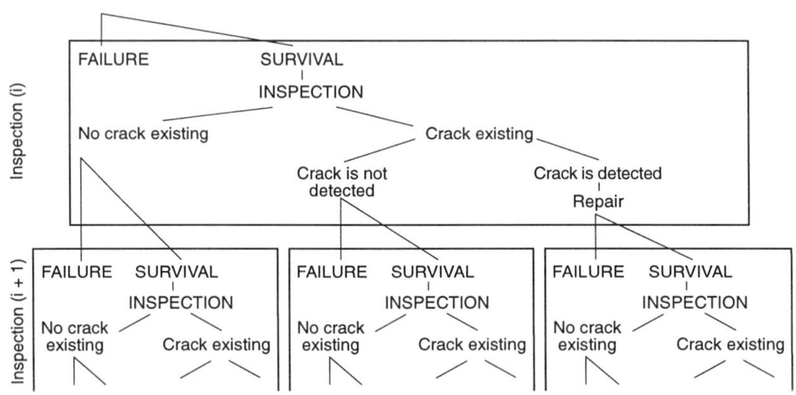

Fig. 10: Inspection scheme

The probability P_D for detecting a crack during an inspection depends mainly on the crack length. For visual standard inspections which are performed every 2.5 years, the curve shown in Fig. 11 has been derived on the basis of experience.

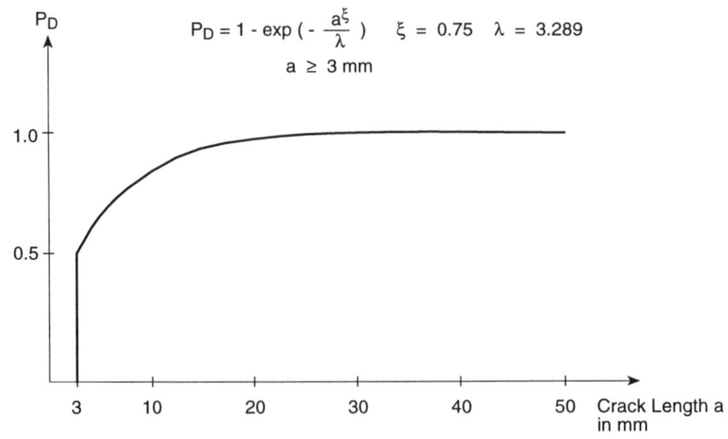

Fig. 11: Probability of detection of a crack vs. crack length

The problem becomes more complex if the crack propagates so slowly that it takes more than one inspection interval before the critical length is reached. The probability for detection P_D or non-detection P_{ND} results from the combination of events considering system reliability techniques. Under consideration of inspections and the probability of detection of cracks, failure of a component occurs,

a) if a crack initiates during an inspection interval and the critical length is reached before the next inspection takes place
b) if the crack propagation is so slow in the beginning that one or more inspection intervals are exceeded and the crack is not detected during any of these inspections.

The probability of failure has been computed in the simulation of crack initiations mentioned, evaluating them according to the above mentioned criteria by the sum of the probability that a crack initiates within an inspection interval i and later becomes critical:

$$P_f = \sum_{i=1}^{n} \left(P_{A,i} \cdot P_{C,i} \right) \tag{10}$$

n = number of inspection intervals

For designs according to the construction rules mentioned [1] (i. e. $P_A = 2.5$ % after 20 years), the probabilities of failure P_f listed in Table 4 are obtained.

Table 4: Computed mean values of number of cycles N_C between crack initiation and reaching the critical crack length a_c and probability of failure P_f

r	50 mm	150 mm	500 mm	500 mm
Y (a/r)	Fig. 3	Fig. 3	Fig. 3	Eq. (6)
N_C	$1.675 \cdot 10^7$	$0.498 \cdot 10^7$	$0.343 \cdot 10^7$	$0.474 \cdot 10^7$
a_c	128.0 mm	36.6 mm	21.2 mm	28.7 mm
P_f	$5.689 \cdot 10^{-4}$	$6.889 \cdot 10^{-3}$	$1.033 \cdot 10^{-2}$	$7.090 \cdot 10^{-3}$

The results show that for the largest notch radius P_f is 20 times higher than for the smallest radius. This effect and the absolute values of P_f are not acceptable.

CONCLUSIONS FOR PERMISSIBLE STRESS RANGES WITH REGARD TO A CONSTANT PROBABILITY OF FAILURE

In order to arrive at the same probability of failure for all notches considered, the permissible stress range or the detail category $\Delta\sigma_R$ according to eq. (2) has to be reduced for increasing notch radius, if the inspection intervals remain unchanged. If the results for r = 50 mm are chosen as a basis, the results summarized in Table 5 are obtained.

Table 5: Computed permissible detail categories $\Delta\sigma_R$ for constant probability of failure P_f

r	50 mm	150 mm	500 mm	500 mm
Y (a/r)	Fig. 3	Fig. 3	Fig. 3	Eq. (6)
$\Delta\sigma_R(r)$ [N/mm^2]	140.0	126.8	123.5	126.3
P_f	$5.689 \cdot 10^{-4}$	$5.689 \cdot 10^{-4}$	$5.689 \cdot 10^{-4}$	$5.689 \cdot 10^{-4}$

The reduction in fatigue strength with increasing notch radius can be approximated by the following equation:

$$\Delta\sigma_R(r) = \left[0.9 + (5/r)\right]\,\Delta\sigma_R \qquad (11)$$
$$50\text{ mm} \le r \le 500\text{ mm}$$

where $\Delta\sigma_R$ = detail category of plate edges acc. to construction rules

The findings of this investigation have been considered in the recently revised construction rules of Germanischer Lloyd [1] which are similar to the new IIW recommendations [7]. This means that notches having the same notch stress are now assessed in [1] in a more or less stringent way depending on their notch radius. This does not necessarily mean that notches with smaller radii are less prone to fatigue - although this holds true for circular holes in large plates. Other notches at plate edges, such as re-entrant or hatch corners, show higher stress concentration with decreasing radius so that smaller radii are, of course, more critical here - even when considering above correction. The application is illustrated by examples in [8].

ACKNOWLEDGEMENTS

The investigation has been financially funded by the German Ministry for Education, Science, Research and Technology.

REFERENCES

1. Germanischer Lloyd (1997): *Rules for Classification and Construction, I - Ship Technology, Part 1: Seagoing Ships, Chapter 1: Hull Structures.* Hamburg.
2. Schwalbe, K.-H. (1980): *Fracture Mechanics of Metallic Materials (in German).* Carl Hanser Verlag, München, Wien.
3. Fricke, W. and Lormes, H. (1993): *Systematic Investigation of alternative solutions for structural details at hatch corners of container ships (in German).* FDS-Report 244/1993, Forschungszentrum des Deutschen Schiffbaus, Hamburg.
4. Haibach, E. (1989): *Fatigue Strength - Procedures and Data for Structural Analysis (in German).* VDI-Verlag GmbH, Düsseldorf.
5. NN. (1991): *COMREL - COMponental RELiability, STRUREL - STRuctural RELiability Analysis Programs.* RCP - Reliability Consulting Programs, München.
6. Almar-Naess, A. Ed. (1985): *Fatigue Handbock - Offshore Steel Structures.* Tapir Publ., Trondheim, 1985.
7. International Institute of Welding (1996): *Fatigue Design of Welded Joints and Components.* Abington Publishing, Cambridge.
8. Fricke, W., Petershagen, H. and Paetzold, H. (1998): *Fatigue Strength of Ship Structures - Examples.* GL-Technology 1/1998, Germanischer Lloyd, Hamburg.

EFFECTS OF INITIAL CRACKS AND FIRING ENVIRONMENT ON CANNON FATIGUE LIFE

J. H. UNDERWOOD and M. J. AUDINO
US Army Armament Research, Development and Engineering Center
Benet Laboratories, Watervliet, NY 12189 USA

ABSTRACT

A case-study description is given of laboratory fatigue life tests of a US Army 155 mm inner diameter cannon tube, performed in the early 1990s. Measured fatigue lives and results from stress and fracture mechanics analyses are used to determine the effects of service conditions on the safe fatigue life of the cannon tube. Fatigue failure in the laboratory tests occurred at nearly the same number of load cycles at the tube inner diameter and at a notch on the tube outer diameter, so the different effects on life at the two locations are considered.

The description of the life test results and the related mechanics analyses include: measured initial crack sizes for different firing environment and analysis of the effect of initial crack size on life; solid mechanics calculations of local applied and residual stresses at the locations of fatigue crack growth; fracture mechanics assessment of fatigue life including effects of initial crack size, applied firing stresses, residual overstrain stresses, and stress concentrations; a log normal statistical analysis of safe fatigue life for various combinations of test results.

The tests and analyses, combined with other related work, show that [i] the use of fatigue intensity factor in a stress - life plot gives a consistent description of fatigue life over a broad range of test variables, including cylinder configuration, initial crack size, applied and residual loading, and material yield strength; [ii] the fatigue intensity factor method can be used to differentiate between fatigue scatter and abnormal fatigue life behavior and to focus on the cause of the abnormal behavior; [iii] a larger than expected initial crack size of 0.05 mm was identified by metallography and found to be the cause of a significant decrease in calculated log-normal safe fatigue life.

KEYWORDS

Fatigue crack growth, cannon, fracture mechanics, residual stress, pressure vessels, high strength steels, stress - life curves, heat-affected crack growth

INTRODUCTION

Recently a new 'fatigue intensity factor' method [1] for representing fatigue life results has been proposed that accounts for two fundamentally important control variables for fatigue life - local stress range and initial crack size - in a single parameter. The new method was shown to give a good description of fatigue life for the extensive test results from high strength steel cannon pressure vessels of Davidson and coworkers [2] and other recent results. The fatigue intensity factor method gave a consistent description of fatigue life for a wide range of vessel configurations and related fatigue failure locations and for steels in the yield strength range of 1000 to 1300 MPa.

The objectives here are [i] to describe the basis of the fatigue intensity factor method of fatigue life assessment using an extensive series of life results from laboratory tests of cannon pressure vessels, and [ii] to determine the key factors that control fatigue life in a recent cannon pressure vessel fatigue case study using the fatigue intensity factor method and the earlier series of laboratory test results. The recent case study is of particular interest because it showed fatigue failures to occur at nearly equal numbers of fatigue cycles at both the smooth inner diameter (ID) surface of the cannon pressure vessel and at the notched outer diameter (OD) surface. Thus, one group of cannon pressure vessel tests included two important types of fatigue failure encountered in pressure vessels. In addition, this group of tests involved both compressive and tensile residual stresses and very different types and sizes of initial crack, all important controlling parameters in fatigue life tests. This wide variety of test conditions and results in one group of tests should result in an informative case study.

ANALYSIS

The basis for the fatigue intensity factor method is the Paris law [3], which describes a significant portion of the fatigue crack growth behavior, da/dN, of metals:

$$da/dN = C\,(\Delta K)^m \tag{1}$$

where ΔK is the positive range of stress intensity factor, C and m are experimental constants, and for steels m is often about 3. Using the classic expression $K = \sigma\,(\pi a)^{1/2}$, an approximate K for small cracks in a variety of configurations with applied tensile stress, σ, and integrating over the range from the initial crack size, a_i, to the critical crack size, a_c, gives:

$$N = [1/(C\,\pi^{m/2}\,\{1 - m/2\}\{\Delta\sigma\}^m)]\,[\,a_c^{(1-m/2)} - a_i^{(1-m/2)}] \tag{2}$$

Taking logs, and using the observation that typically $a_c >> a_i$, leads to:

$$\log(\Delta\sigma \times a_i^{(1/2-1/m)}) = (-1/m)\log N - (1/m)\log\{(m/2 - 1)\,C\,\pi^{m/2}\} \tag{3}$$

which can be recognized as a straight line on log coordinates with slope (-1/m) and intercept - (1/m) log $\{(m/2 - 1)\,C\,\pi^{m/2}\}$ which are constant for a given material. Equation (3) suggests that plots of log $(\Delta\sigma \times a_i^{(1/2-1/m)})$ versus log N will fall on a single straight line with (-1/m) slope and that all the critical stress range and initial crack size information will be included in the single parameter $(\Delta\sigma \times a_i^{(1/2-1/m)})$, which becomes $(\Delta\sigma \times a_i^{1/6})$ for m = 3.

Finally, the effect of material yield strength is empirically added to the single parameter representation of stress range and initial crack size, to obtain the fatigue intensity factor:

$$\text{fatigue intensity factor} = \Delta\sigma \times (S_{y\text{-ave}} / S_y) \times a_i^{1/6} \quad (4)$$

where the ($S_{y\text{-ave}}/S_y$) term effectively increases the stress range for a specimen with yield strength, S_y, lower than the mean value, $S_{y\text{-ave}}$, and decreases the stress range for a yield strength higher than mean. In the latter case for example, a decreased stress range corresponds to a higher fatigue life, as is often observed for an increase in material strength. Equation (4) is the parameter proposed to describe the intensity of the fatigue loading of a structural component, including the important effects of stresses and initial crack at the failure site, as well as the effect of variation of material strength within a group of tests. The use of this parameter to describe fatigue life over a wide range of test conditions is shown in Figure 1, for the twelve series of fatigue life tests listed in Table 1. Hydraulic oil pressurization of 1-2 m long sections of thirty nine cannon tubes of forged A723 pressure vessel steel were performed over a period on many years. Failure locations in the twelve test series include the ID surface of the tube, the ID surface of a hole through the tube wall, and the fillet radius of a notch on the tube OD.

Table 1. Test conditions for pressure vessel fatigue life tests

Location/	replicates	Yield Strength (MPa)	Inner Radius (mm)	Outer Radius (mm)	Applied Pressure (MPa)	Residual Stress (MPa)	Stress Range (MPa)	Fatigue Life (cycles)	Initial Crack (mm)
ID:	3	1280	89	187	345	0	1275	10,094	0.01
	3	1200	89	187	345	0	1275	10,039	0.1
	4	1270	89	187	345	0	1275	4,110	0.5
	6	1020	89	187	345	-680	758	23,152	0.01
	4	1120	79	155	670	-711	1247	5,590	0.1
	3	1230	89	142	393	-546	896	10,629	0.1
HOLE:	2	1240	53	76	207	0	1797	5,240	0.01
	2	1170	60	94	297	0	1657	5,535	0.01
	2	1220	78	107	83	0	664	42,025	0.01
NOTCH:	3	1230	78	142	393	+397	1196	11,960	0.01
	2	1240	85	153	406	+ 31	1397	10,605	0.01
	5	1070	60	135	670	-530	1702	3,159	0.01

Note in Table 1 that the test conditions and parameters have considerable variation over the twelve series of laboratory fatigue tests. Most tests had stress concentration at the failure site, due to rifling at the ID or due to a hole or notch. For the four series in which there was a compressive residual stress present at the failure site, this was accounted for in calculating the stress range, as described later. The only parameter that is not explicitly measured or calculated is the initial crack size. For eight of the twelve test series the initial crack size at the unaffected machined surface where fatigue failure initiated was determined to be 0.01 mm, based on metallographic investigation of the steel and the machining processes used for the vessels. Note that the behaviour of very short initial cracks that has been considered in recent years is not included in the analyses here. For the four test series in which the failure site was exposed to the cannon firing

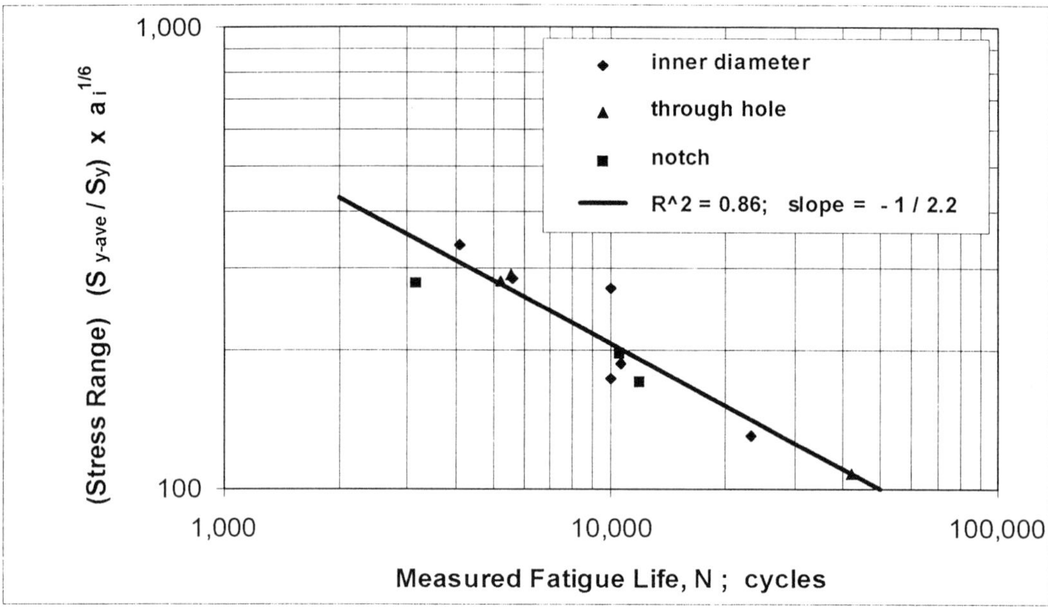

Fig. 1. Fatigue life results from twelve series of high strength steel cannon pressure vessel tests plotted versus fatigue intensity factor

environment before the laboratory test, the depth of heat-check cracks produced by firing, determined metallographically, was used: 0.1 mm for three series, and an unusually deep 0.5 mm for one series.

Figure 1 shows a logarithmic plot of the average measured fatigue life from the twelve test series versus the fatigue intensity factor determined from Eq. (4) and the information in Table 1. The linear regression line through the data has a reasonably high R^2 correlation coefficient, 0.86, and the slope, -1/2.2, is in reasonable agreement with the ideal slope of -1/3 for m = 3, discussed earlier. Considering the wide range of pressure vessel configuration, test conditions, and failure locations, this fatigue intensity factor description of fatigue life is considered to be quite useful for life assessment. Prior work [1] showed that description of fatigue life based only on stress range resulted in much poorer correlation. The remainder of the discussion here will use the fatigue intensity factor method and the results of Figure 1 in a case study evaluation of a recent series of cannon pressure vessel tests.

CANNON PRESSURE VESSEL CASE STUDY

Fatigue Tests

The breech end of seven cannon tubes were used for hydraulic pressure tests to determine the fatigue life of the cannon. The section of cannon tube tested had a stepped inner radius of 89 mm at one end and 78 mm at the other, and a constant outer radius of 142 mm. The steel used was a forged ASTM A723 pressure vessel steel heat treated to a nominal 1230 MPa yield strength. Measured values of yield strength,

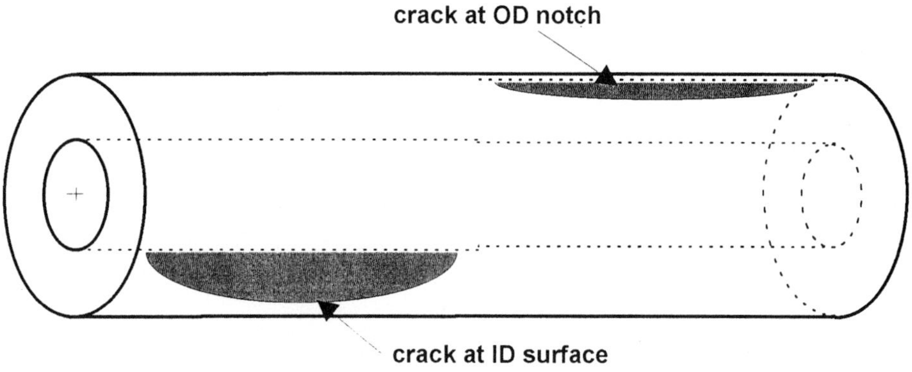

Fig. 2. Two locations of fatigue cracking encountered in case study

Charpy impact energy and fracture toughness from the seven tubes are listed in Table 2. Cyclic pressurization of 393 MPa pressure was applied at a frequency of about 1 Hz. The resulting fatigue failures occurred at two quite different locations: at the ID surface of the 89 mm inner radius portion of the test specimen where the initial crack depth was 0.1 mm; and at an OD notch present in the 78 mm inner radius portion of the test specimen, where the initial crack depth in the notch root was expected to be 0.01 mm. See Figure 2 and Table 2. The OD notch was added after overstraining, to prevent rotation of the cannon tube during firing. Photographs of the fracture surfaces were obtained after the test. Figure 3 shows the fracture surface of specimen #01, which failed from the OD notch. The bottom, lightly shaded portion of the fracture surface is the fatigue crack emanating from the notch root, the middle portion is a relatively flat part of the fast fracture region, and the top, darker portion is the so-called shear-lip portion of fast fracture progressing to the ID surface . Figure 4 shows the fracture surface of specimen #02, which failed from the ID surface. The classic semi-elliptical shape of a surface crack is clearly seen, with a small region of shear-lip as the crack approached the OD surface at the top of the photo. Failure at two remote locations at about the same fatigue life indicates that, by critical design or good fortune, the local stresses at the two different locations where fatigue cracking started are of similar magnitude. The analysis of local stress range at the two failure sites is considered next.

Table 2. Cannon pressure vessel case study tests

Specimen Number	Yield Strength (MPa)	Impact Energy (J)	Fracture Toughness (MPa $m^{1/2}$)	Failure Location —	Initial Crack (mm)	Fatigue Life (cycles)
# 01	1214	30	134	OD notch	0.01	13,800
# 02	1235	32	189	ID surface	0.1	10,319
# 03	1228	32	152	ID surface	0.1	13,067
# 05	1228	33	151	OD notch	0.01	10,828
# 09	1249	30	135	OD notch	0.01	11,252
# 11	1244	29	113	OD notch	0.01	5,501
# 25	1230	20	131	ID surface	0.1	8,501

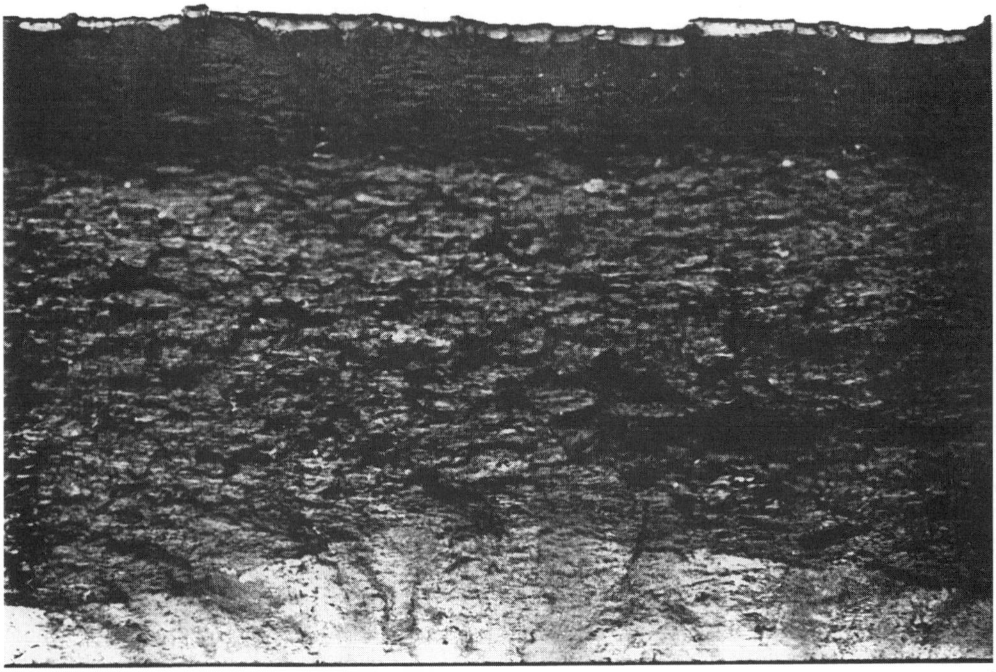

Fig. 3 Fracture surface of Specimen #01; OD notch failure; 1.5 magnification

Fig. 4 Fracture surface of Specimen #02; ID surface failure; 0.5 magnification

Local Stress Range at Failure Sites

Expressions for the local positive effective stress range, $\Delta\sigma$, at the two locations where fatigue crack growth was observed are as follows:

$$\Delta\sigma_{ID} = \sigma_\theta + \sigma_{ov} + p_{crack} \tag{5}$$

$$\Delta\sigma_{OD} = k_t\,\sigma_\theta - \sigma_r \tag{6}$$

The expression in Eq. 5 for the ID cracking includes the stresses that often have the primary control of fatigue cracking in a pressure vessel, the applied and overstrain residual hoop direction stresses at the ID surface, σ_θ and $\sigma_{ov.}$ As is common for pressure vessels, these tubes were overstrained, in this case to the extent that plastic deformation proceeded to about 60% through the tube wall thickness. The stress range in Eq. 5 is considerably reduced by accounting for the overstrain residual stress; no effect of residual stress on mean stress is considered here. The third term accounts for the pressure in the crack that produces the equivalent of a tensile stress oriented normal to the crack plane that is equal in magnitude to the applied pressure.

The expression in Eq. 6 for the cracking at the OD notch includes the applied hoop direction stresses at the depth below the OD corresponding to the notch root, where the crack initiated. The notch depth, width and fillet radius are 8, 13 and 1.5 mm , respectively, resulting in a stress concentration factor, k_t, of 3.26, significantly increasing the local stress range at the crack site. The second term in Eq. 6 accounts for the compressive radial direction stress, σ_r, that effectively adds a small amount to the hoop stress at the notch. An addition to the usual tensile hoop stress is made, equal to the (negative) value of the compressive radial stress at the notch root location. Standard expressions are available for calculating σ_θ and σ_r [4] and σ_{ov} [5]. Note, however, that a 30% reduction in σ_{ov} compared to the standard linear unloading calculation was used here, to account for the Bauschinger reduction in ID residual stress for an overstrained tube [6].

The values of local stress range from Eqs. 5 and 6 are used in Eq. 4 to determine fatigue intensity factor, which is used to assess fatigue life, considered next.

Fatigue Life Analysis of Case Study Results

The seven case study fatigue test results are shown on a plot of fatigue intensity factor versus measured fatigue life, see Figure 5. The prior fatigue test results discussed in relation to Figure 1 are also shown, as is the linear regression for the prior results. Six of the seven case study results are grouped reasonably well around the regression line, whereas the result for specimen #11, with a fatigue life of 5501 cycles, lies significantly away from the other results. As might be expected, the significantly lower fatigue life of specimen #11 caused some concern. One possible reason that was considered for the lower life of specimen #11 was inferior material properties, but this was considered unlikely because only one of the three properties for specimen #11 listed in Table 2 was noticeably lower, and only slightly lower than expected. Attention was focused on initial crack size. As shown in Table 2 the initial crack size for specimen #11 was taken as 0.01 mm, the value shown in prior work to be typical of machined surfaces of A723 steel that have not been affected by the firing environment. However, metallographic investigation of the notch area of specimen #11 showed a 0.05 mm thick layer of apparently heat-affected material at the notch root, where the fatigue crack growth started. Since the OD notch was not subjected to any firing damage, it was determined that the observed heat-affected layer was caused by a rapid machining process. Recalculation

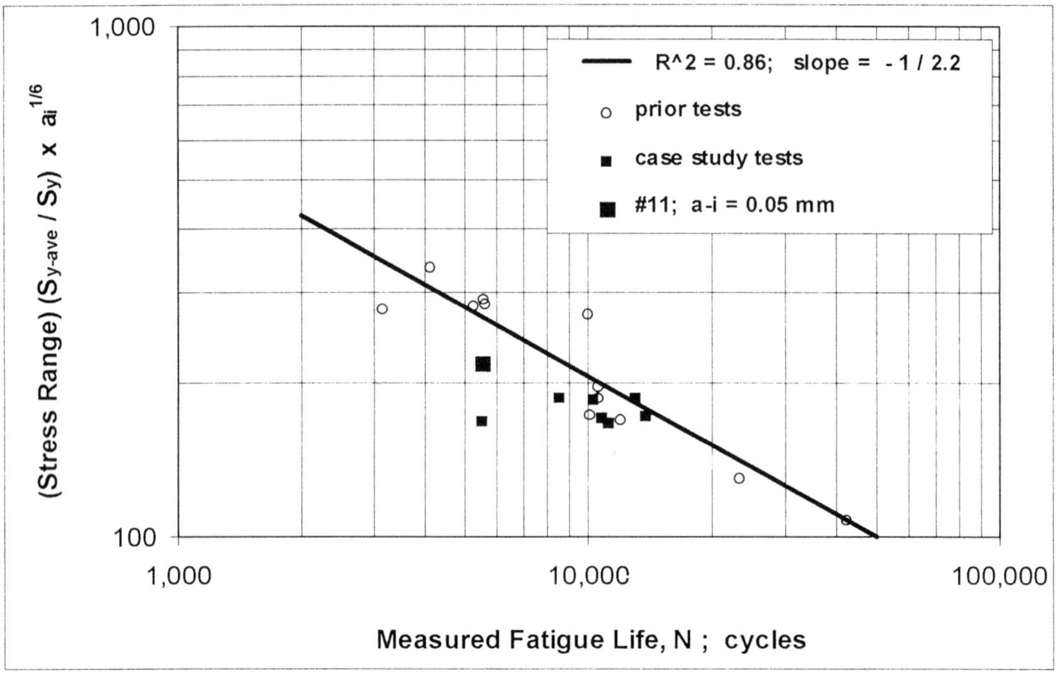

Fig. 5 Fatigue life results from the case study tests compared with prior results

of fatigue intensity factor using a_i = 0.05 mm for specimen #11, rather than 0.01 mm, resulted in much improved agreement of the specimen #11 data with the other data; see Figure 5.

Additional physical understanding of the importance of the larger than anticipated initial crack size of specimen #11 on fatigue life can be obtained from the Paris equation [3], discussed earlier. A simple integrated form of Eq. 3 [7] provides an approximate calculation of life that will, at least, show the relative effect of variation in initial crack size, a_i on calculated life, N_{cacl}. The calculation is:

$$N_{cacl} = 2 / a_i^{1/2} \, C \, (1.99 \, k_t \, \sigma_\theta)^3 \qquad [7]$$

In Eq. 7 the constant 2 is from integration; C is the constant from Eq. 3, 6.52 x 10^{-12} for the steels used in cannon tubes; the constant 1.99 is appropriate for a short edge crack; k_t is 3.26 as discussed earlier; and σ_θ is the applied hoop stress at the notch root location, 361 MPa $m^{1/2}$ in this case. Using these values in Eq. 7, the calculated lives for a_i = 0.01 and 0.05 mm, the originally expected and actual measured initial crack lengths for specimen #11, respectively, are: 7,400 and 3,300 cycles. The calculated lives are very nearly in the same ratio as the measured lives: 11,960 cycles (mean of #01, #05, #09) compared with 5,501 cycles for specimen #11. Also, considering the simplicity of Eq. 7, the absolute agreement of the calculated and measured lives is remarkable.

The implications of the specimen #11 result on the calculation of a safe fatigue life using a log-normal statistical analysis is summarized in Table 3. Analysis is shown for various combinations of the seven fatigue life results of the case study, using the one-sided normal distribution tolerance factors available in statistics texts, such as [8]. An expression for a safe life, N*, can be written in terms of the mean and standard deviation of the natural log of the lives in a given sample, as follows:

$$N^* = exp[\ (ln\ N)_{AVE}\ - k\ (ln\ N)_{SD}] \qquad [8]$$

Table 3. Log-normal statistical analysis of pressure vessel fatigue lives

Sample Size and Makeup --	Mean Life (cycles)	Standard Deviation (cycles)	Tolerance Factor —	Life for 90% Confidence and 0.99 Reliability N*; (cycles)
5; 01-09	11,853	1,502	4.67	6,569
6; 01-11	10,795	2,921	4.24	2,557
7; 01-25	10,467	2,804	3.97	2,936
6; 01-09, 25	11,295	1,918	4.24	5,337

where k is the normal distribution tolerance factor [8] for a given confidence that at least a certain proportion of a population will be above the safe life. Table 3 shows safe lives for 90% confidence and 0.99 reliability for various samples of the test results. If the first five test results of Table 2 are used, a safe life of 6,569 cycles is obtained. Then if specimen #11 with the lower than expected life is added to the sample, the safe life is dramatically reduced, to 2,557 cycles. Only slight improvement is obtained by adding the seventh test result to the sample. The use of these significantly lower safe lives is the correct approach for a sample of test results that may include the lower than expected fatigue life of specimen #11. However, if there is assurance that all specimens such as #11 are removed from the sample, such as by screening all specimens and removing those which were rapidly machined, then the higher safe life of 5,337 cycles given at the bottom of the table can be used. It should be noted that in rigorous application of safe life calculation procedures, fatigue lives from tests with two different failure locations would not be combined in a single calculation. Because of the close agreement of lives, once specimen #11 was excluded, test results with different failure locations were combined in this case study.

DISCUSSION OF RESULTS AND SUMMARY

The fatigue intensity factor approach to fatigue life assessment uses fracture mechanics to obtain a single parameter representation of stress range, initial crack size and material yield strength. Use of this single parameter in a stress - life plot of fatigue results gives a consistent description of fatigue life over a broad range of test variables, including: cylinder configuration; initial crack size as affected by firing environment; applied pressure and overstrain residual stress; and material yield strength.

A single set of recent fatigue results with two very different failure locations and applied and residual stresses compared well with the prior results using the fatigue intensity factor approach. The stresses and stress concentrations that contribute to the local stress range and control fatigue cracking were outlined. A fatigue intensity factor plot of the recent results showed a clear outlier. Investigation showed that the outlier had five times the expected initial crack size, and that when the larger initial crack was accounted for, the agreement with other results was restored. Thus, the fatigue intensity factor method can be used to identify the presence of an outlier from the central trend of fatigue life results and focus on the cause of abnormal fatigue life behavior.

An outlier with significantly lower life than the mean was shown to have a dramatic effect on calculated log-normal safe life. Including the test which had a life 49% below the mean life of five other tests caused a 61% reduction in calculated safe life. Thus, unless it can be shown that all specimens with the larger than expected initial crack size are removed from the group of specimens intended for service, there will be a severe effect on the safe service life of the pressure vessels in the case study.

REFERENCES

1. Underwood, J. H. and Parker, A. P. (1997) In: *Advances in Fracture Research,* B. L. Karihaloo, Y-W Mai, M. I. Ripley and R. O. Ritchie (Eds.) Pergamon, Amsterdam, pp. 215-226.

2. Davidson, T. E., Throop, J. F. and Underwood, J. H. (1977) In: *Case Studies in Fracture Mechanics*, T. P. Rich and D. J. Cartwright (Eds.) US Army Materials and Mechanics Research Center, Watertown, MA, pp. 3.9.1 - 3.9.13.

3. Paris, P. C. and Erdogan, F. (1963) *J. of Basic Engineering*, **85**, 528-534.

4. Roark, R. J. and Young, W. C. (1975) *Formulas for Stress and Strain,* McGraw-Hill, New York

5. Hill, R. (1950) *The Mathematical Theory of Plasticity*, Clarendon Press, Oxford.

6. Parker, A. P. and Underwood, J. H. (1998) In: *Fatigue and Fracture Mechanics: 29th Volume, ASTM STP 1332*, T. L. Panontin and S. D. Sheppard (Eds.)American Society for Testing and Materials, West Conshohocken, PA.

7. Parker, A. P., Underwood, J. H., Throop, J. F. and Andrasic, C. P. (1983) In: *Fracture Mechanics: Fourteenth Symposium - Volume I; Theory and Analysis, ASTM STP 791,* J. C. Lewis and G. Sines (Eds) American Society for Testing and Materials, Philadelphia, pp. 216-237.

8. Burlington, R. S. and May, D. C., Jr. (1970) *Handbook of Probability and Statistics with Tables,* McGraw-Hill, New York, p. 394.

WEIGHT FUNCTIONS AND STRESS INTENSITY FACTORS FOR EMBEDDED CRACKS SUBJECTED TO ARBITRARY MODE I STRESS FIELDS

G. GLINKA and W. REINHARDT
Department of Mechanical Eng., University of Waterloo, Waterloo, Ontario, Canada N2L 3G1

ABSTRACT

Fatigue cracks in welded, shot peened and case hardened machine components are subjected to various stress fields induced by the load and the residual stress. Both stress field types are highly non-linear and appropriate handbook stress intensity solutions are unavailable for such configurations. The method presented below is based on the generalized weight function technique enabling the stress intensity factors to be calculated for planar cracks subjected to any Mode I stress field. Both the general weight functions and the calculated stress intensity factors are validated against numerical and analytical data. The numerical procedure for calculating stress intensity factors for arbitrary non-linear stress distributions is briefly discussed as well. The method is particularly suitable for modeling fatigue crack growth of single buried elliptical and multiple cracks.

KEYWORDS

Stress intensity factor, weight function, non-linear stress field

NOTATION

a - depth of a semi-elliptical, elliptical (minor semi-axis) or edge crack
A - the deepest point of surface, semi-elliptical crack
B - the surface point of semi-elliptical crack
c - half length of semi-elliptical or elliptical crack (major semi-axis)
G_c – crack contour
G_b – external free boundary contour
K_I - mode I stress intensity factor (general)
K_{IA} - mode I stress intensity factor at the deepest point A
K_{IB} mode I stress intensity factor at the surface point B
M_i - coefficients of weight functions (i= 1, 2, 3)
M_{iA} - coefficients of the weight functions for the deepest point A (i= 1, 2, 3)
M_{iB} - coefficients of the weight functions for the surface point B (i= 1, 2, 3)
m(x,a) - weight function (general)
$m_A(x,a)$ - weight function for the deepest point A of a semi-elliptical surface crack
$m_B(x,a)$ - weight function for the surface point B of a semi-elliptical crack
Q - elliptical crack shape factor
S - external (applied) load
SIF – stress intensity factor
s – the shortest distance between the load point and the crack contour

t – thickness
Γ_c – inverted crack contour
Γ_b – inverted external free boundary contour
Φ - angle co-ordinate for parametric representation of an ellipse
Ω - crack area
ρ - distance between the point load and any point A on the crack front
$\sigma(x)$ - a stress distribution over the crack surfaces
σ_0 - nominal or reference stress (usually the maximum value of $\sigma(x)$)
x - the local, through the thickness co-ordinate
Y - geometric stress intensity correction factor

INTRODUCTION

Fatigue durability, damage tolerance and strength evaluation of cracked structural components require calculation of stress intensity factors for cracks located in regions characterized by complex stress fields. This is particularly true for cracks emanating form notches or other stress concentration regions that are frequently found in engineering practice. Such components require fatigue analysis of crack propagating through a variety of interacting stress fields. Moreover, these are often planar two-dimensional surface or buried cracks with irregular shapes. The existing handbook stress intensity factor solutions are not sufficient in such cases due to the fact that most of them have been derived for simple geometry and load configurations. The variety of notch and crack configurations, and the complexity of stress fields occurring in engineering components require more versatile tools for calculating stress intensity factors than the currently available ready made solutions, obtained for a range of specific geometry and load combinations.
Therefore, a method for calculating stress intensity factors for one- and two-dimensional cracks subjected to two-dimensional stress fields is discussed below. The method is based on the use of the weight function technique.

STRESS INTENSITY FACTORS AND WEIGHT FUNCTIONS

Most of the existing methods of calculating stress intensity factors require separate analysis of each load and geometry configuration. Fortunately, the weight function method developed by Bueckner [1] and Rice [2] simplifies considerably the determination of stress intensity factors. The important feature of the weight function is that it depends only on the geometry of the cracked body. If the weight function is known for a given cracked body, the stress intensity factor due to any load system applied to the body can be determined by using the same weight function. The success of the weight function technique for calculating stress intensity factors lies in the possibility of using superposition. It can be shown, [3], that the stress intensity factor for a cracked body (Fig. 1) subjected to the external loading, S, is the same as the stress intensity factor in a geometrically identical body with the local stress field $\sigma(x)$ applied to the crack faces. The local stress field, $\sigma(x)$, induced in the prospective crack plane, is determined for *uncracked* body that makes the stress analysis relatively simple. Therefore, if the weight function is known there is no need to derive ready made stress intensity factor expressions for each load system and associated internal stress distribution. The stress intensity factor for a one dimensional crack can be obtained by multiplying the weight function, m(x,a), and the internal stress distribution, $\sigma(x)$, in the prospective crack plane, and integrating the product along the crack length 'a'.

$$K = \int_0^a \sigma(x)m(x,a)dx \tag{1}$$

The weight function, m(x,a), can be interpreted (Fig.2) as the stress intensity factor that results from a pair of splitting forces, P, applied to the crack face at position x.

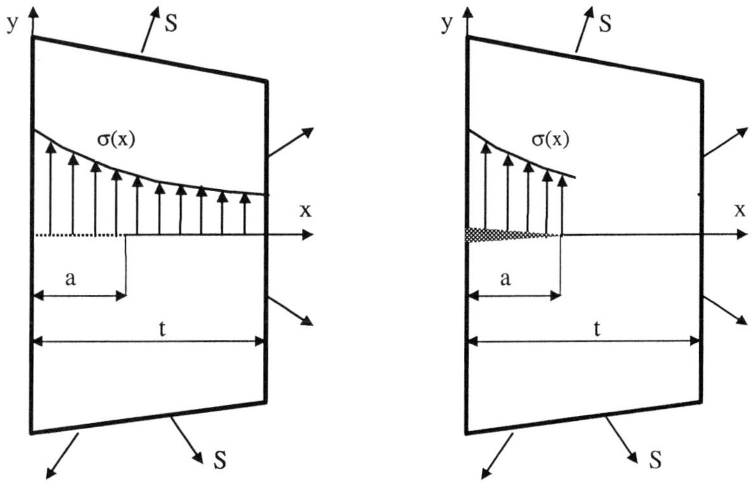

Fig.1. Nomenclature and the concept of superposition

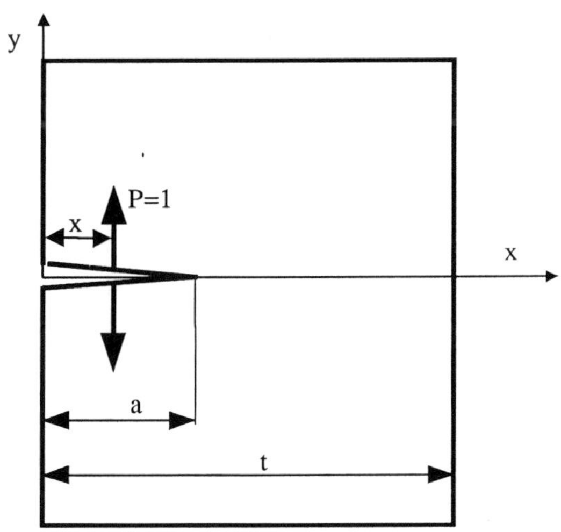

Fig. 2. Weight function for an edge crack in a finite width plate; nomenclature

Since the stress intensity factors are linearly dependent on the applied loads, the contributions from multiple splitting forces applied along the crack surface can be superposed and the resultant stress intensity factor can be calculated as the sum of all individual load contributions. This results in the integral, (1), of the product of the weight function, m(x,a), and the stress function, σ(x), for a continuously distributed stress field. A variety of one-dimensional (line-load) weight functions can be found in references [4,5,6]. However, their mathematical forms vary from case to case and therefore they are not easy to use. Therefore, Shen and Glinka [7] have proposed one general weight function expression, which can be used for a wide variety of Mode I cracks.

UNIVERSAL WEIGHT FUNCTIONS FOR ONE-DIMENSIONAL STRESS FIELDS

The weight function is dependent on the geometry only and in principle should be derived individually for each geometrical configuration. However, Glinka and Shen [7] have found that one general weight function expression can be used to approximate weight functions for a variety of geometrical crack configurations subjected to one-dimensional stress fields of Mode I.

$$m(x,a)=\frac{2}{\sqrt{2\pi(a-x)}}\left[1+M_1\left(1-\frac{x}{a}\right)^{\frac{1}{2}}+M_2\left(1-\frac{x}{a}\right)^{1}+M_3\left(1-\frac{x}{a}\right)^{\frac{3}{2}}\right] \quad (2)$$

As an example the system of coordinates and the notation for an edge crack are given in Fig. 2. In order to determine the weight function, m(x,a), for a particular cracked body, it is sufficient to determine, [8], the three parameters M_1, M_2, and M_3 in expression (2). Because the mathematical form of the weight function, (2), is the same for all cracks, the same methods can be used for the determination of parameters M_1, M_2, and M_3 and for the integration routine for calculating stress intensity factors from eq.(1). The method of finding the M_i parameters has been discussed in reference [8].

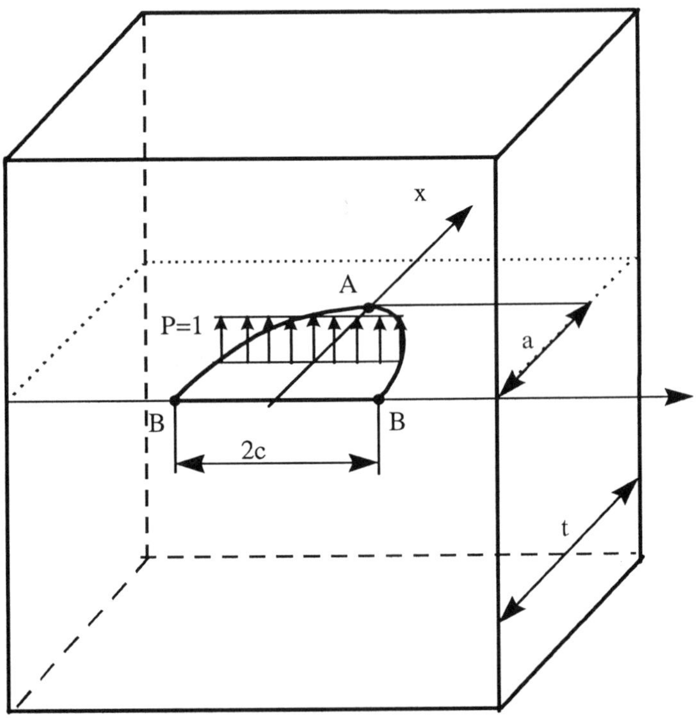

Fig. 3. Semi-elliptical surface crack under the unit line load; weight function notations

Moreover, it has been found that only limited number of generic weight functions is needed to enable the calculation of stress intensity factors for a large number of load and geometry configurations. In the case of 2-D cracks such as the surface breaking semi-elliptical crack in a finite thickness plate or cylinder, the stress intensity factor changes along the crack front. However, in many practical cases the deepest point, A, and the surface point, B, are associated (Fig. 3) with the highest and the lowest value of the stress intensity factor respectively.

Therefore, weight functions for the points A and B of a semi-elliptical crack have been derived, [9], analogously to the universal weight function of eq.(2).

- For point A (Fig. 3)

$$m_A(x,a,a/c,a/t) = \frac{2}{\sqrt{2\pi(a-x)}}\left[1 + M_{1A}\left(1-\frac{x}{a}\right)^{\frac{1}{2}} + M_{2A}\left(1-\frac{x}{a}\right)^{1} + M_{3A}\left(1-\frac{x}{a}\right)^{\frac{3}{2}}\right] \tag{3}$$

- For point B (Fig. 3)

$$m_B(x,a,a/c,a/t) = \frac{2}{\sqrt{\pi x}}\left[1 + M_{1B}\left(\frac{x}{a}\right)^{\frac{1}{2}} + M_{2B}\left(\frac{x}{a}\right)^{1} + M_{3B}\left(\frac{x}{a}\right)^{\frac{3}{2}}\right] \tag{4}$$

The weight functions, $m_A(x,a)$ and $m_B(x,a)$, for the deepest and the surface points, A and B, respectively have been derived for the crack face unit *line loading* making it possible to analyze one-dimensional stress fields (Fig. 3), dependent on one variable , x , only. A variety of universal line load weight functions [9-13] have been derived and published already.
In order to calculate stress intensity factors using the weight function technique the following tasks need to be carried out:

- Determine stress distribution, $\sigma(x)$, in the prospective crack plane using linear elastic analysis of uncracked body (Fig. 1a), i.e. perform the stress analysis ignoring the crack and determine the stress distribution $\sigma(x) = \sigma_0 f(S,x)$;
- Apply the "uncracked" stress distribution, $\sigma(x)$, to the crack surfaces (Fig. 1b) as traction
- Choose appropriate generic weight function
- Integrate the product of the stress function $\sigma(x)$ and the weight function, m(x,a), over the entire crack length or crack surface, eq.(1).

WEIGHT FUNCTIONS FOR TWO-DIMENSIONAL STRESS FIELDS

In spite of the efficiency and great usefulness of the line load weight functions, they cannot be used if the stress field is of two-dimensional nature, i.e. where the stress field, $\sigma(x,y)$, in the crack plan depends on the x and y coordinates. Therefore in order to calculate stress intensity factors for planar cracks of arbitrary shape subjected to two-dimensional stress field weight functions for a point load (Fig. 4) are needed. A two-dimensional point-load weight function, $m_A(x,y)$, represents the stress intensity factor at point, A, on the crack front (Fig. 4), induced by a pair of forces, P, attached to the crack surface at point P(x,y). If the weight function is given in a closed mathematical form, it makes it possible to calculate the stress intensity factor at any point along the crack front. In order to determine the stress intensity factor induced by a two-dimensional stress field, $\sigma(x,y)$, at a point, A, on the crack front the product of the stress field, $\sigma(x,y)$, and the weight function, $m_A(x,y)$, needs to be integrated over the entire crack surface area Ω.

$$K_A = \iint_{\Omega} \sigma(x,y) m_A(x,y;P)\,dxdy \tag{5}$$

Rice has shown [14] that the 2-D point load weight function for an arbitrary planar crack in an infinite body can be generally written as:

$$K_A = m_A(x,y,P) = \frac{P\sqrt{2s}}{\pi^{3/2}\rho^2} w(x,y;P) \tag{6}$$

The function w(x,y;P) accounting for the effect of the crack geometry is usually unknown and it has to be determined for each particular crack geometry.

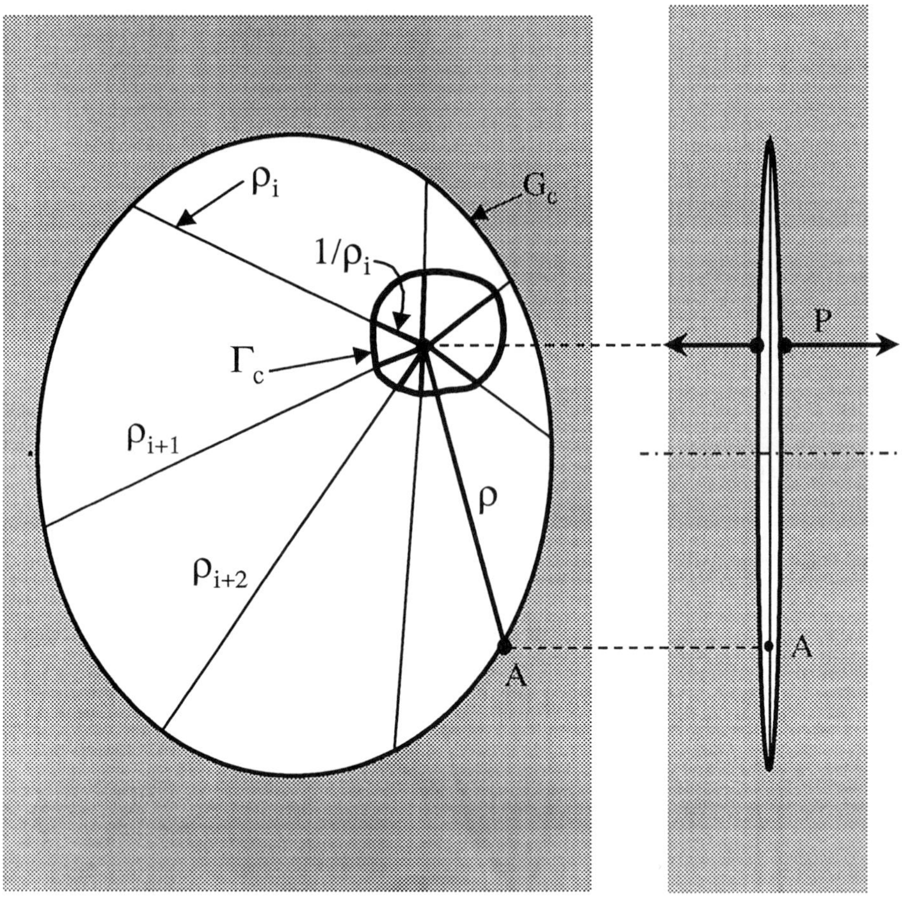

Fig. 4. Notation for the 2-D weight function

Oore and Burns [15] proposed a general 2-D weight function (7) from each the function w(x,y;P) can be derived for a few known crack shapes.

$$K_A = m_A(x,y;P) = \frac{P\sqrt{2}}{\pi\rho^2\sqrt{\oint\limits_{G_c}\frac{dG_c}{\rho^2}}} \tag{7}$$

The notation for the weight function (7) is given in Fig. 4. Oore and Burns have shown [15] that after deriving closed form expressions for the line integral in equation (7) several exact weight functions could be derived for straight and circular cracks in infinite bodies. However, some difficulties were encountered concerning integration of the weight function. In its original form the weight function (7) was also unable to account for finite boundaries of a cracked body. Therefore, Oore and Burns [15] proposed a simplified boundary correction routine for a limited number of geometry and load configurations.

However, it has been found that the integration procedure could be significantly simplified if one uses a geometrical interpretation of the line integral in expression (7). Namely, it can be proved that the line integral $\oint_{G_c} \frac{d\,G_c}{\rho^2}$ represents the arc length, Γ_c, of the crack contour inverted (Fig. 4) with respect to the point, P(x,y), where the load P is applied. As a consequence the weight function (7) can be written in a simpler and easier for integration form.

$$K_A = m_A(x,y,P) = \frac{P\sqrt{2}}{\pi\rho^2\sqrt{\Gamma_c}} \tag{8}$$

The inverted contour can also be looked at as the locus of inverted radii $1/\rho_i$. It can be further proved that the inverted contours form circles in the case of infinite straight line and circular contours. Therefore, the general weight function (8) makes it possible to derive closed form weight functions for a variety of straight and circular crack configurations.

ELLIPTICAL CRACK SUBJECTED TO TWO-DIMENSIONAL STRESS FIELD

In order to verify the 3-D capability of the weight function (8) the SIFs were calculated for a an elliptical crack in an infinite body subjected to 2-D stress field (9) shown in Fig. 5.

$$\sigma(x,y) = \sigma_0 \frac{xy}{ac} \tag{9}$$

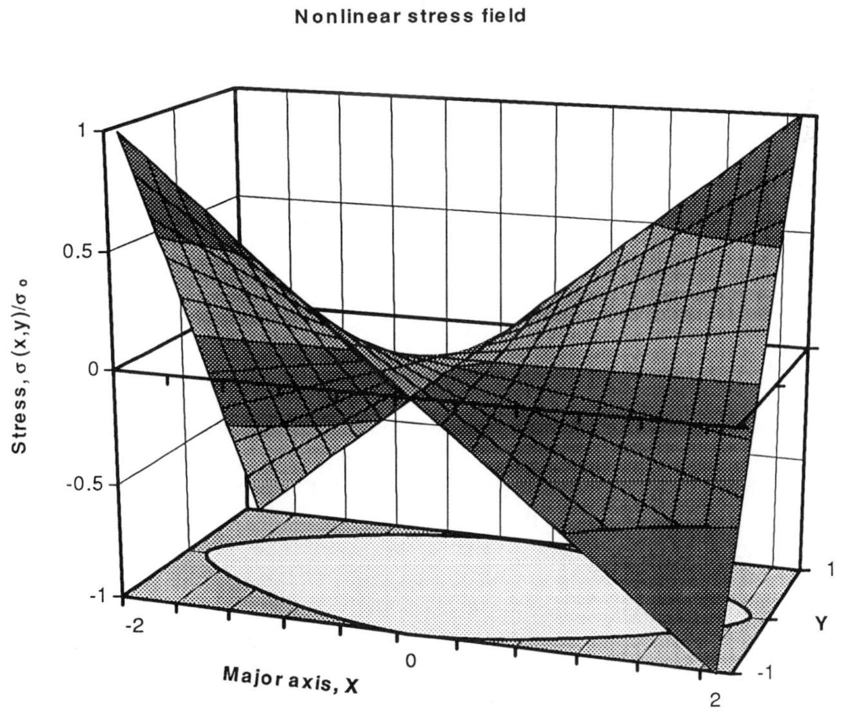

Fig. 5. Two-dimensional nonlinear stress field applied to the elliptical crack

The stress intensity factors were determined for several points on the crack front defined by the parametric angle **Φ.** The agreement between the weight function based SIFs and the data obtained by Shah and Kobayashi [16] was very good for a wide range of ellipse aspect ratios a/c. The data points shown in Fig. 6 were obtained for a/c=0.2. Good agreement between the weight function based SIF and Kobayashi's [16] data indicates that both the shape of the crack and the stress distributions were adequately accounted for.

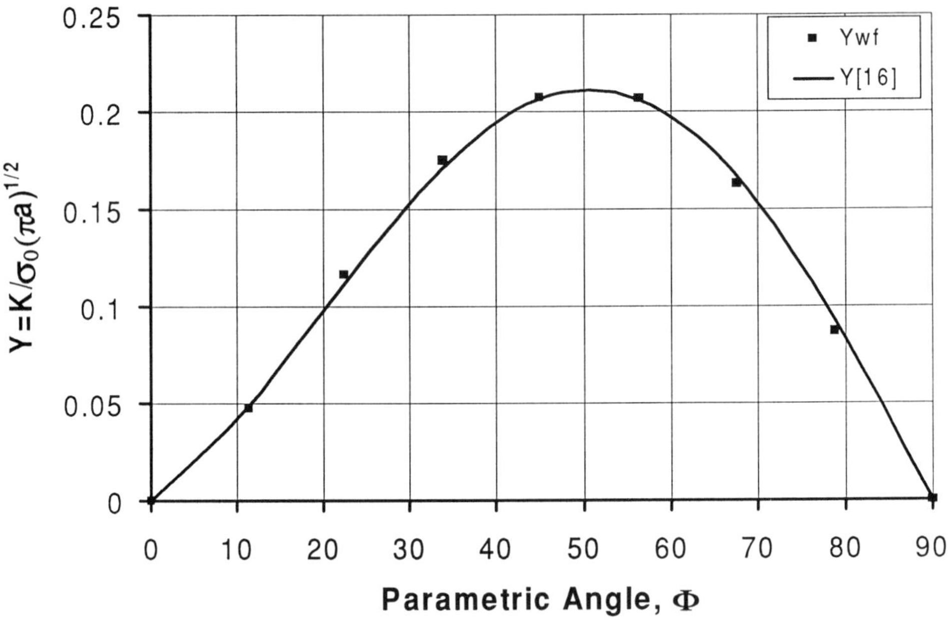

Fig. 6. Comparison of SIF values with Kobayashi's data [16] obtained for an elliptical crack subjected to 2-D stress field, a/c=0.2

CRACKS IN FINITE BODIES

The example presented above indicates that the general point load weight function (8) supplies very accurate SIF results for cracks in infinite bodies. However, in the case of finite bodies both the crack contour and the free boundary contour have to be accounted for. The influence of these two boundaries on the stress intensity factor at a point on the crack front is not the same in nature. The increase (expansion) of the crack boundary increases the stress intensity factor while the increase of the free boundary decreases the SIF.

An analysis of the existing [4] weight functions in finite bodies led to the conclusion that the effect of the free boundary depends on the location of the point on the crack contour with respect to the external free boundary contour. This effect can be expressed in terms of an analogous inverted arc as in the case of the inverted crack contour. The free boundary contour is inverted with respect to the point on the crack front where the SIF is going to be determined. The general 2-D weight function accounting for the free boundary effect was found to be:

$$K_A = m_A(x,y;P) = \frac{P\sqrt{2}}{\pi\rho^2} \times \frac{\sqrt{\Gamma_c + \Gamma_b}}{\Gamma_c} \tag{10}$$

The notation and geometrical interpretation of the inverted arcs are illustrated in Fig. 7.

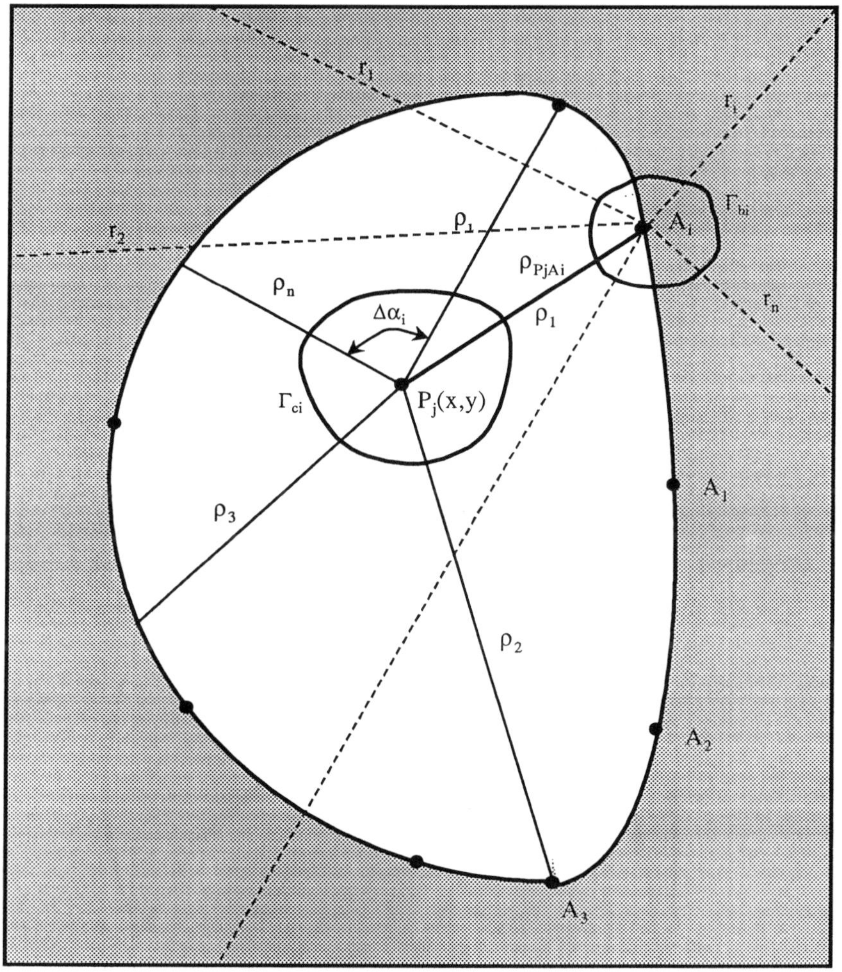

Fig. 7. Notation for the generalized point load weight function for an arbitrary planar crack in a finite body

There are very few SIF solutions for elliptical cracks in finite bodies subjected to complex 2-D stress distributions. Raju & Newman [17] derived the first solution for embedded elliptical crack in a finite thickness plate subjected to uniform tensile stress by using FEM method. Two crack configurations shown in Fig. 8 were analyzed here. In both cases the weight function SIFs were no more than a few percent different from the FEM data. The distributions of SIFs along the crack contour are presented in Fig. 9. The parametric angle was assumed to be $\Phi=0$ at the end of the semi-minor axis and $\Phi=\pi/2$ at the end of the semi-major axis. The agreement is good for both cracks regardless of their location with respect to the free boundaries.

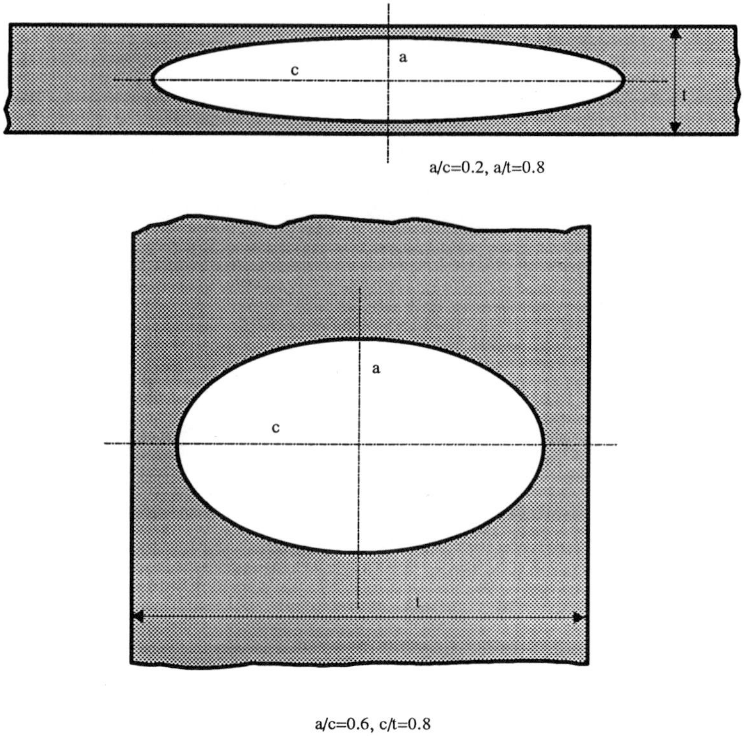

Fig. 8. Elliptical crack in a finite thickness/width plate

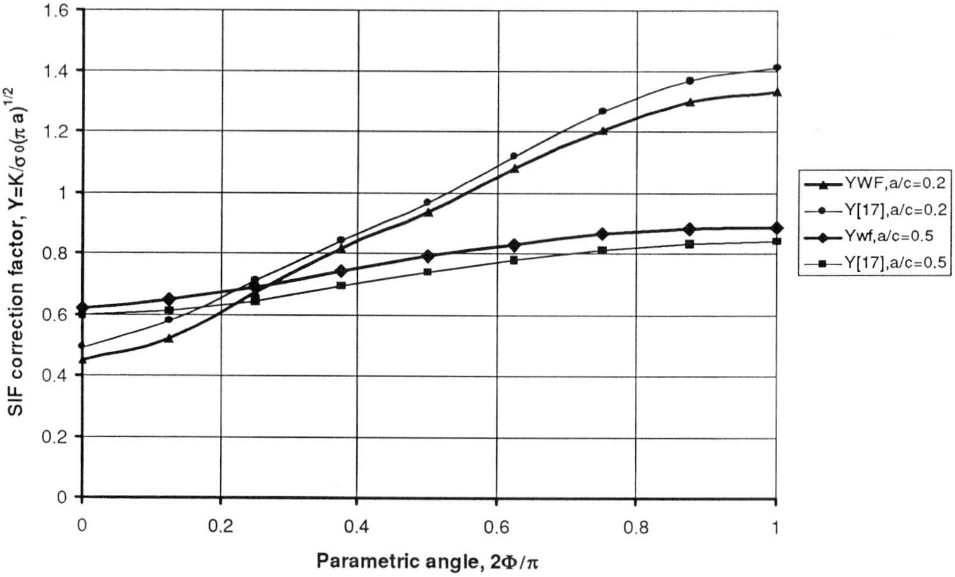

Fig. 9. Comparison of the weight function based SIF with FEM data of Raju & Newman [17]

NUMERICAL TECHNIQUE

The stress intensity factor due to a continuous stress field applied to the crack surface is calculated by integrating the product (5) of the weight function and the stress field over the entire crack area. In the case of numerical calculation the continuous stress field has to be replaced by a finite number of forces, P_j.

$$P_j = \sigma(x_j, y_j) \times \Delta\Omega_j \tag{11}$$

where: $\Delta\Omega$ - is the surface area of the crack segment over which the stress $\sigma(x_j,y_j)$ is acting.
The stress intensity factor induced by such a force (Fig. 7) is equal to the value of the weight function associated with that point. After substitution of eq.(11) into (10) we can write the weight function in terms of stresses applied to the crack surface.

$$m_A(x_A, y_A; P_j) = \frac{\sigma(x_j, y_j) \times \Delta\Omega_j \times \sqrt{2}}{\pi\rho^2_{P_jA}} \times \frac{\sqrt{\Gamma_{c,j} + \Gamma_{b,i}}}{\Gamma_{c,j}} \tag{12}$$

The stress intensity factor K_A at point A on the crack contour induced by stress field $\sigma(x,y)$ has to be calculated finally as the sum of contributions from all crack area segments $\Delta\Omega_j$.

$$K_A = \sum_{j=1}^{j=N} \left\{ \frac{\sigma(x_j, y_j) \times \Delta\Omega_j \times \sqrt{2}}{\pi\rho^2_{P_jA}} \times \frac{\sqrt{\Gamma_{c,j} + \Gamma_{b,i}}}{\Gamma_{c,j}} \right\} \tag{13}$$

The dimensions of crack surface segments, $\Delta\Omega_j$, have to be small in order to approximate sufficiently accurate the stress gradient and the crack geometry. In order to calculate stress intensity factor using the weight function technique described above the following tasks need to be carried out.

- Determine the stress distribution, $\sigma(x,y)$, in the prospective crack plane using linear elastic analysis of uncracked body, i.e. perform the stress analysis ignoring the crack (Fig. 1) and determine the stress distribution $\sigma(x,y) = \sigma_0 * f(S,x,y)$;
- Apply the "uncracked" stress distribution, $\sigma(x,y)$, to the crack surfaces as traction.
- Choose point A on the crack front where the SIF is to be calculated (Fig. 7).
- Calculate the length of the inverted external boundary contour Γ_b associated with point A on the crack front.
- Calculate the length of the inverted crack contour, Γ_{cj}, associated with the point load at location, $P_j(x_j,y_j)$, where the stress, $\sigma_j(x_j,y_j)$, is acting.
- Calculate the stress intensity factor contribution (the weight function) due to the stress, $\sigma_j(x_j,y_j)$, applied at point, $P_j(x_j,y_j)$, on the crack surface (eq. 12)Repeat the calculations for point A over the entire crack surface area substituting appropriate values for stress $\sigma_j(x_j,y_j)$, the length of arc Γ_{cj} and add all the contributions (eq. 13).

CONCLUSIONS

The proposed method has been applied to a variety of crack geometry and load configurations. The weight function and the numerical integration procedure make it theoretically possible to calculate stress intensity factor for any convex planar crack subjected to any Mode I stress distribution. The comparison with exact and numerical SIF data has revealed that the methodology discussed above simulated correctly the SIF distribution along the crack contour for several geometry and load configurations. In majority of cases analyzed up to date the agreement with analytical and numerical data was within 5% of error.

REFERENCES

1. Bueckner, H. F., (1970), A novel principle for the computation of stress intensity factors. *Zeitschrift fur Angewandte Mathematik und Mechanik*, vol. 50, pp.529-546
2. Rice J. R., (1972), Some remarks on elastic crack-tip stress field, *International Journal of Solids and Structures,* vol. 8, pp. 751-758
3. Broek, D., (1988), *The Practical Use of Fracture Mechanics*, Amsterdam,Kluwer
4. Tada, H., Paris, P., and Irwin, G., (1985), *The Stress Analysis of Cracks Handbook*, 2nd Edn, Paris Production Inc., St. Louis, Missouri, USA
5. Wu, X., R., and Carlsson, A., J., (1991), *Weight Functions and Stress Intensity Factor Solutions*, Pergamon Press, Oxford, UK
6. Fett, T. and Munz, D., (1994), *Stress Intensity Factors and Weight Functions for One-dimensional Cracks*, Report No. KfK 5290, Kernforschungszentrum Karlsruhe, Institut fur Materialforschung, Dezember 1994, Karlsruhe, Germany
7. Glinka, G. and Shen, G., (1991), Universal features of weight functions for cracks in mode I, *Engineering Fracture Mechanics*, vol. 40, pp. 1135-1146,
8. Shen, G. and Glinka, G., (1991), Determination of weight functions from reference stress intensity factors, *Theoretical and Applied Fracture Mechanics*, vol. 15, pp. 237-245,
9. Shen, G. and Glinka, G., (1991), Weight Functions for a Surface Semi-Elliptical Crack in a Finite Thickness Plate, *Theor. Appl. Fract. Mech.*, Vol. 15, No. 2, pp. 247-255
10. Zheng, X.J., Glinka, G. and Dubey, R., (1995), Calculation of stress intensity factors for semi-elliptical cracks in a thick-wall cylinder, *Int. J. Pres. Ves, & Piping*, vol. 62, pp 249- 258
11. Zheng, X.J., Glinka, G. and Dubey, R., (1996), Stress intensity factors and weight functions for a corner crack in a finite thickness plate, *Engineering Fracture Mechanics*, vol. 54, No. 1, pp 49-62
12. Wang, X. and Lambert, S. B., (1995), Stress intensity factors for low aspect ratio semi-elliptical surface cracks in finite-thickness plates subjected to non-uniform stresses, *Engineering Fracture Mechanics*, vol. 51, pp.517-532
13. Wang, X. and Lambert, S. B., (1998), Stress intensity factors and weight functions for high aspect ratio semi-elliptical surface cracks in finite-thickness plates, *Engineering Fracture Mechanics*, (to be published)
14. Rice, J., *Weight Function Theory for Three-dimensional Elastic Crack Analysis,* in Fracture Mechanics Perspectives and Directions (20th Symposium), ASTM STP 1020, American Society for Testing and Materials, 1989, pp.29-57
15. Oore, M and Burns, D.J.,(1980), Estimation of stress intensity factors for embedded irregular cracks subjected to arbitrary normal stress fields, *Jrnl of Pressure Vessel Technology, ASME*, vol.102, pp 202-211
16. Shah, R.C. and Kobayashi, A.C., (1971), Stress intensity factor for an elliptical crack under arbitrary normal loading, *Engineering Fracture Mechanics*, vol. 15, pp. 71-96
17. Newman, J. C. and Raju, I. S., *Stress-Intensity Factor Equations for Cracks in Three-Dimensional Finite Bodies Subjected to Tension and Bending Loads*, in Computational Methods in the Mechanics of Fracture, Ed. S. N. Atluri, North-Holland,1986, pp. 311-334

A MODIFIED FRACTURE-MECHANICS METHOD FOR THE PREDICTION OF FATIGUE FAILURE FROM STRESS CONCENTRATIONS IN ENGINEERING COMPONENTS

D.TAYLOR
Mechanical Engineering Dept., Trinity College, Dublin, Ireland

ABSTRACT

This paper describes a new method, known as 'crack modelling', which is designed for the prediction of fatigue behaviour in components for which elastic stress field data are available through finite element (FE) analysis or similar methods. The approach is based on the well-known result that sharp notches can be modelled as cracks, allowing an equivalent stress-intensity factor to be calculated. The fatigue limit of a sharply-notched specimen is thus found by equating the equivalent stress intensity factor range to the material's threshold value. This approach has been extended notched specimens of standard geometry under simple applied loads to components of arbitrary geometry with complex loading modes. This paper illustrates the application of the method through a number of case studies of components. The reliability of the method is then assessed in two ways. Firstly, the sensitivity of the predictions to errors in the FE analysis and the processing of FE data is examined. Secondly, the issue of conservatism in the prediction is discussed; predictions suitable for engineering design should ideally be conservative, though the degree of conservatism should not be too great. It is concluded that the method is a robust one which is not strongly dependant on accurate FE meshing and which normally contains a useful degree of conservatism, though further work is needed in the treatment of physically small notches.

KEYWORDS

Fatigue, finite element analysis, fracture mechanics, stress concentrations, reliability.

INTRODUCTION

Fatigue failure in engineering components and structures almost invariably occurs at stress concentrations, such as geometric features, defects, joints or material discontinuities. The use of finite element (FE) analysis and other numerical methods can give a very good picture of local stresses at these so-called 'hot-spots'. However, we are unable to make the best use of this data due to inadequacies in the theory of fatigue analysis for stress concentrators, especially for sharp notches and for low-strength materials. The analysis of fatigue at notches generally takes one of three forms. Firstly, stress-life approaches simply correlate the local maximum stress at the stress-concentration (the 'hot-spot stress') with a stress-life (S/N) curve for the material. Normally a linear-elastic FE analysis is used, and this can lead to huge amounts of conservatism in the prediction of fatigue limit, for example, because plasticity is not accounted for. Secondly, strain-life methods (usually called the 'local strain approach') avoid this problem to some extent by accounting for plasticity, but these methods require more complex analysis (elastic-plastic FE) and access to more detailed material data such as cyclic stress-strain curves. Even so, considerable amounts of conservatism can occur. Thirdly, fracture mechanics (FM) methods have been used, attempting to describe the growth of a crack from the hot-spot. These methods have had limited success to date because of the difficulties of estimating the rate of short-crack growth within the microstructure and due to problems associated with the varying elastic-plastic stress field. Even more that the strain-life method, FM assessments require a large amount of special knowledge of material properties.

Therefore a method is needed which can accurately describe fatigue failure initiated from hot-spots, which is capable of taking into account complex component geometry and which requires only simple elastic FE analysis and readily-available material property data. The method known as "crack modelling" has been developed to meet these requirements. In what follows we will consider only high-cycle fatigue, and limit the discussion to the prediction of fatigue limit or fatigue strength: the cyclic stress below which failure will not occur, within a specified number of cycles. This is the primary requirement in most industrial design problems.

The crack modelling method is an extension of the theory described by Smith and Miller (1), which was developed to predict the fatigue limit of specimens containing notches of known depth, D, root radius, ρ and stress concentration factor K_t. These workers noted that for blunt notches, when K_t is less than some critical value which we can call K_t^*, the simple stress-life approach works well; i.e. failure occurs in the notched specimen if the hot-spot stress range is greater than the material's normal fatigue limit, $\Delta\sigma_o$, as measured from plain specimens, at the same R ratio. For sharp notches ($K_t > K_t^*$), this method becomes increasingly conservative and unusable, but Smith and Miller noted that an accurate prediction could be obtained simply by modelling the notch as a crack of the same depth, D. Thus a fracture mechanics equation of the form:

$$\Delta K_{th} = F\Delta\sigma_{on}(\pi D)^{1/2} \qquad (1)$$

...can be used to predict the fatigue limit of the notched specimen, $\Delta\sigma_{on}$ (expressed here as the applied nominal stress) in terms of the notch depth, D and the material's crack-propagation threshold stress intensity range, ΔK_{th}. The factor F accounts for other geometry and loading factors. This approach has been assessed by Taylor and O'Donnell, and shown to give good predictions of fatigue limit for a wide range of materials and notch geometries (2). However, it can only be used on notches of standard geometry, for which D is known and for which equation 1 is valid with a known value for F. This greatly limits its practical use; a way is needed for extending the method to allow it to consider stress concentrators of any geometry. This has been achieved using the crack modelling method, which is summarised in fig.1. The stress field for the component under consideration is obtained using FE analysis: elastic stresses only are obtained. This is compared with the stress field for a crack of standard geometry, known as the model crack - in this case the model used is a centre crack of length $2a_w$ under uniform applied tension σ_w, but other geometries could conceivably be used instead. The values of the model parameters (a_w and σ_w) are adjusted to give the smallest possible difference between the two stress fields, which are defined in terms of a stress-distance curve: a plot of stress along a line drawn from the hot spot. The stress used is the maximum principal stress and the direction of the line is normal to the direction of the maximum principal stress at the hot spot. It is assumed that (a) a crack will grow initially along this line, driven by the maximum principal stress, and (b) its threshold for growth (and therefore the fatigue limit of the component) will occur when the ΔK value of the model crack is equal to ΔK_{th} for the material. Essentially we propose that the component, loaded with some arbitrary set of cyclic loads ΔL, will

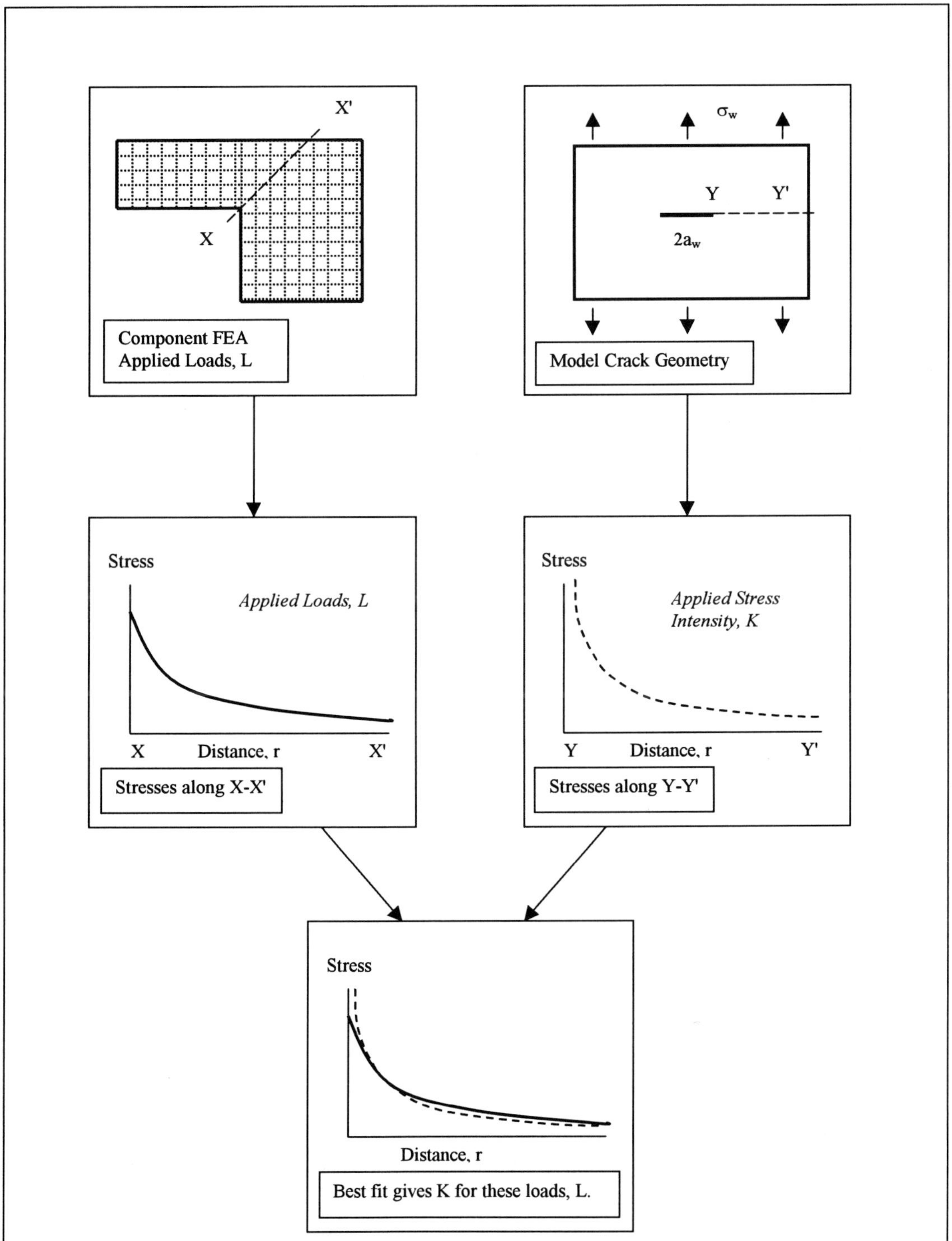

Fig.1 Outline of the crack-modelling technique

behave in the same way as the chosen model crack, even though the geometry of the two situations is very different, because the two stress fields are, approximately, the same.

It has been demonstrated that this method works for the trivial case of standard notch geometries (13), when its predictions are very close to those of Smith and Miller (1). We will now consider its application to two real engineering components.

CASE STUDIES

This section describes two case studies which demonstrate the accuracy of the method as applied to real engineering components.

1) Automotive Crankshaft

Some initial results from the study of this component have been reported previously (3). A crankshaft used in Rover vehicles (fig.2) was tested to failure in the laboratory, using fully reversed (R=-1) cycling. Two types of loading were used: bending (in the plane of the diagram of fig.2) and torsion (twisting about the long-axis of the shaft).

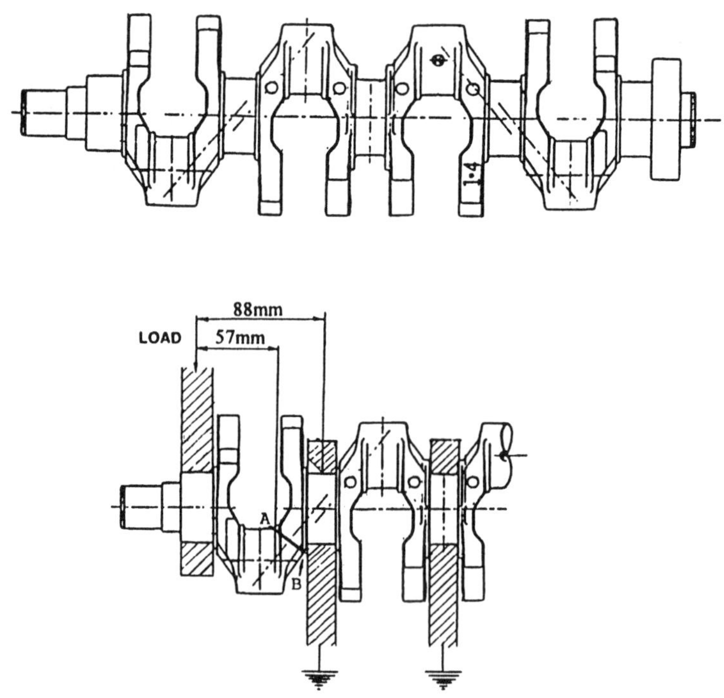

Fig.2. Geometry of the crankshaft component, and its loading during testing.

Table 1 shows the measured fatigue strengths in terms of applied load (or torque) range at ten million cycles to failure. In both cases failure occurred from a fillet radius associated with a bearing; bending loads induced failures normal to the axis of the shaft, initiating at the fillet at point A and growing along A-B. Under torsion the cracks also initiated at the fillet but grew at approximately 45° to the shaft axis. The material used was an SG cast iron which had a fatigue limit (measured using standard tensile specimens) of 580MPa and a threshold ΔK_{th}=23.5MPa(m)$^{1/2}$, at R=-1. A finite element model was constructed using IDEAS software: fig.3 shows stress-distance curves derived from the model for

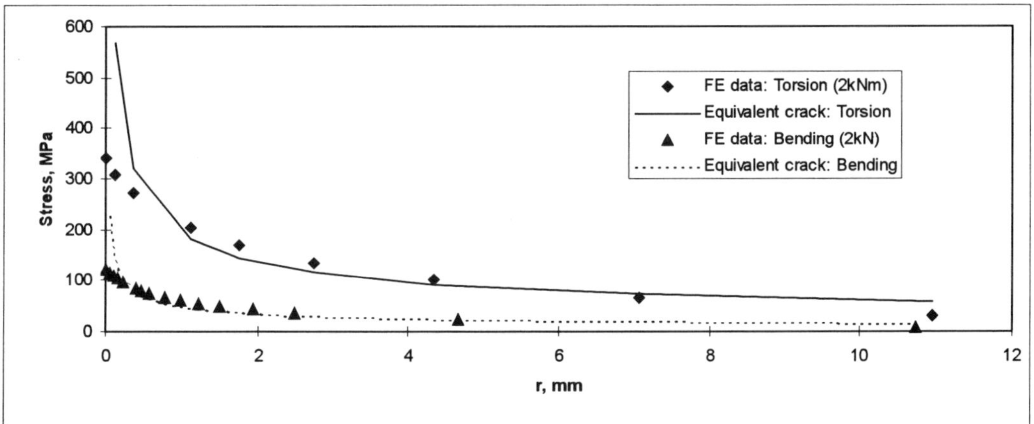

Fig.3. Stress-distance curves for the crankshaft in bending and torsion, comparing FE results and curves estimated for the equivalent model crack.

the two types of loading. Table 1 shows predictions using the crack modelling method. These are very accurate, falling within likely experimental errors for the test results and FE analysis. The method was equally successful in predicting failure from bending and torsion loads.

Table 1. Experimental and Predicted Fatigue Limits for the Crankshaft, expressed in terms of the range of applied load (or torque) at R=-1.

Type of Loading	Experimental Fatigue Strength (at 10^7 cycles)	Predicted Fatigue Strength (%age error)
Bending	12.0kN	12.26kN (2%)
Torsion	3.2kNm	3.10kNm (3%)

2) Marine Component

The analysis of this component has been described in more detail elsewhere (4). This large casting, whose identity cannot be revealed for reasons of confidentiality, experienced fatigue failure in service in its original design and loading configuration. Cracking initiated from a 90° fillet of radius 0.3mm (fig.4). Subsequently two modifications were attempted: in modification A the fillet radius was increased to 3.18mm; in modification B the fillet radius was unchanged but the maximum operating loads were reduced. Fig.5 shows stress-distance curves (deduced from FE modelling such as shown in fig.4) for the three conditions. Modification A clearly reduces the maximum stress considerably, but in practice, failures continued to occur with this design. Modification B showed no failures despite the relatively modest reduction in stress near the hot-spot.

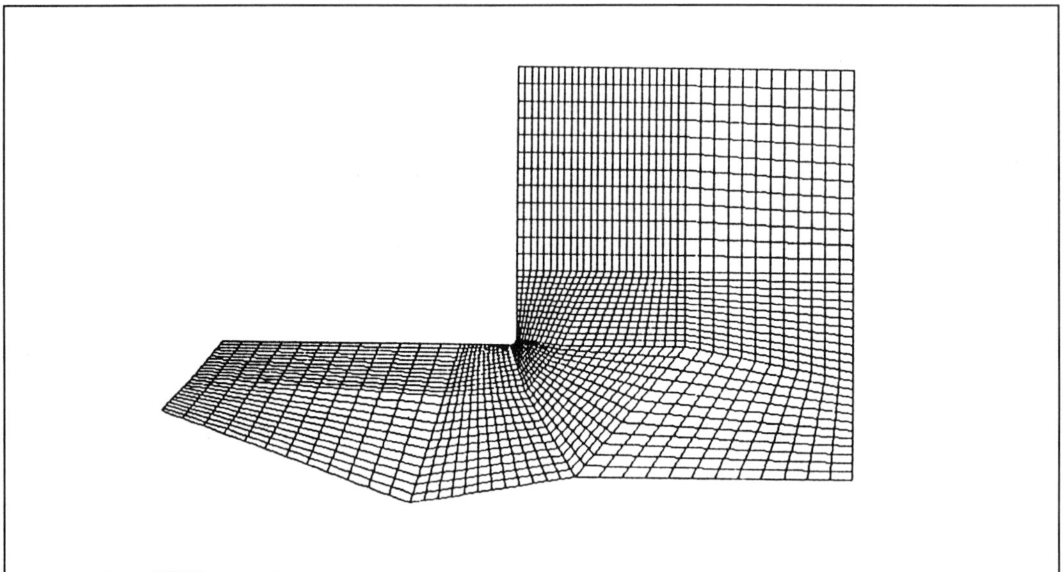

Fig.4. Geometry of the fillet from which failure occurred in the marine component (original design), showing also the FE mesh.

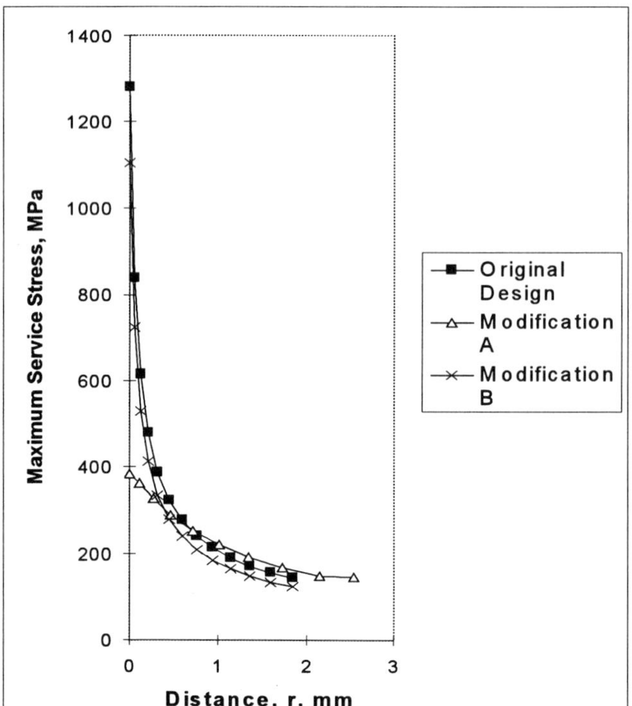

Fig.5. Stress-distance curves for the three designs of the marine component.

Table 2 shows the results of crack modelling analysis which used data on ΔK_{th} measured at various R ratios (4) and compared the relevant threshold value to the estimated cyclic stress intensity for each modification. The analysis correctly predicts the failures of the original design (since $\Delta K > \Delta K_{th}$) and also the non-failures of modification B. In the case of modification A the prediction implies non-failure but the difference between ΔK and ΔK_{th} is very small (8%) - within the range of error of the experimental measurements and FE analysis - so it would be concluded that this modification would be in some danger of failing. More importantly, the method has been able to put the three designs in the correct order of performance, showing that modification B is safer than modification A, despite the great reduction in stress which A seems to achieve.

Table 2. Predictions of the behaviour of the marine component

Design	Estimated ΔK	Estimated R ratio	ΔK_{th} at estimated R ratio	%age difference between ΔK and ΔK_{th}
Original	7.28	0.56	6.83	+6.6%
Modification A (blunt fillet radius)	6.30	0.56	6.83	-7.8%
Modification B (reduced loads)	5.03	0.65	5.93	-15.2%

These case studies have demonstrated that the method can be applied to full-scale components of complex shape, under realistic types of loading. These two examples both involved cast irons, which are traditionally difficult materials to assess because of their low notch-sensitivity and complex stress-strain behaviour. The method has also been tested successfully for other materials, including mild steels (5) and aluminium alloys (6).

RELIABILITY ASSESSMENTS

Mesh Density

FE models used by industrial designers normally have relatively few elements due to the constraints of computer time and space when modelling large complex structures. However, a very fine mesh may be needed in order to obtain accurate stress information for a small, sharp stress concentration. To investigate the sensitivity of the crack modelling technique to mesh density, the same component (the crankshaft) was modelled used two different meshes: a fine mesh (from which the results above were obtained) and a coarser mesh. Element sizes in the region of the hot spot were 0.8mm and 2.8mm for the fine and coarse meshes respectively. Use of the coarse mesh gave a large reduction in the value of the hot spot stress, from 361MPa to 273MPa. By contrast the prediction of the stress intensity, and therefore of the fatigue limit, changed by only 4%, from 12.26kN to 11.76kN. This demonstrates that the method is not sensitive to mesh density.

Limits of the Stress/Distance Curve

In order to carry out the curve-fitting exercise shown in fig.1, in which the stress-distance curves of the component and model crack are compared, some choices have to be made regarding the range of values of distance, r, over which the curves are compared. The minimum value, r_{min}, must be greater than zero, since the crack's stress value rises to infinity at this point. In practice the smallest useful value of r_{min} is generally dictated by the nodal spacings. The maximum value, r_{max}, can be as large as the specimen dimensions dictate, but the model is unable to deal with negative values, so the curve is curtailed if stress values become compressive. These choices are rather arbitrary, so we investigated the effect on the K estimates due to changing these limits. Fig.6 shows the effect on K for the crankshaft loaded in bend with a load of 1kN, which gives a value of $4.06MPa(m)^{1/2}$ when the largest possible range of r values is used. The two graphs show the effect of changing r_{max} whilst keeping r_{min} constant at a low value, and likewise the effect of changing r_{min}. It can be seen that the result is relatively insensitive to the choice of r values: for example, increasing r_{min} to 1mm, which is effectively removing a large amount of information regarding the stress field of this 2mm-radius fillet, changes the K estimate by only 6%. Fig.7 shows similar results for a standard notch (depth 2.5mm, K_t=8.1), loaded in tension, for which the correct analytical result for K is $10.0MPa(m)^{1/2}$.

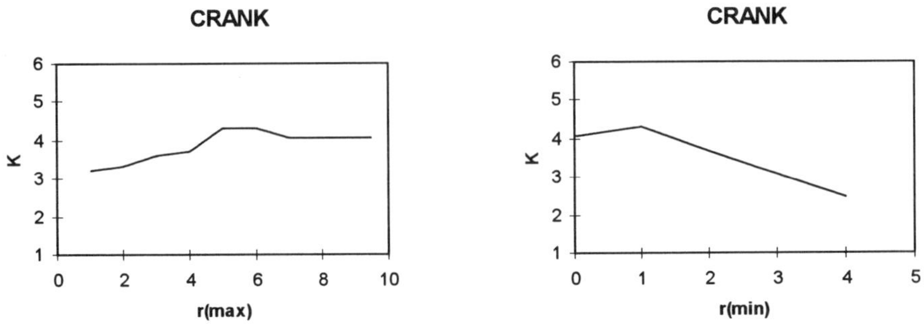

Fig.6. The effect of r_{max} and r_{min} on the estimated K value for the crankshaft. Units of r are mm and of K are $MPa(m)^{1/2}$.

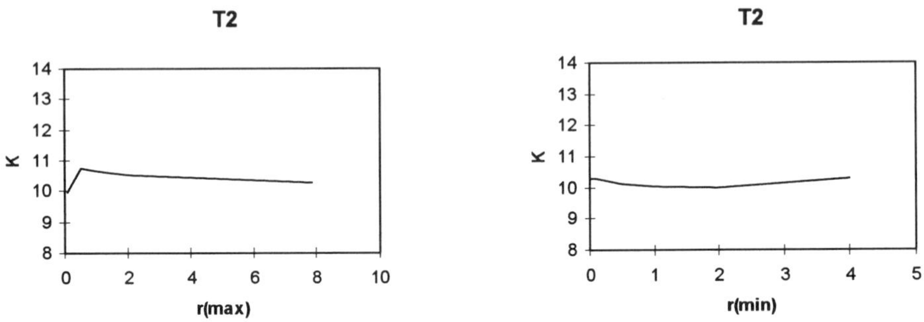

Fig.7. As fig.6, assessing a standard notch loaded in tension.

Size-Effect Errors

It is well known that short cracks display anomalous growth behaviour which is not well predicted by normal fracture mechanics methods (7,8). One way to express this anomaly is the well-known Kitagawa curve, in which the fatigue limit for a specimen containing a crack is plotted as a function of the crack length (fig.8). Using logarithmic scales, the fracture mechanics prediction (corresponding to constant ΔK_{th} in equation 1) gives a straight line, implying that the fatigue limit will approach infinity as the crack length approaches zero. In practice the fatigue limit for zero crack length will be $\Delta\sigma_o$ (the horizontal line in the figure) so as length decreases the fatigue limit values will depart from the fracture mechanics line, tending to the horizontal line at very low lengths. This is substantiated by the experimental data, and ElHaddad et al (9) have suggested a modification to the fracture mechanics equation which provides a reasonable fit to the data:

$$\Delta K_{th} = F\Delta\sigma_{on}(\pi(a+a_o)^{1/2}) \qquad (2)$$

...where a is the crack length and a_o is a material constant.

We can speculate that a similar effect will arise in the case of crack-like notches, and this effect has been demonstrated for notches of simple geometry by Murakami and Endo (10), Peterson (11) and Mitchell (12), who have also proposed empirical correction factors which differ from equation 2. This

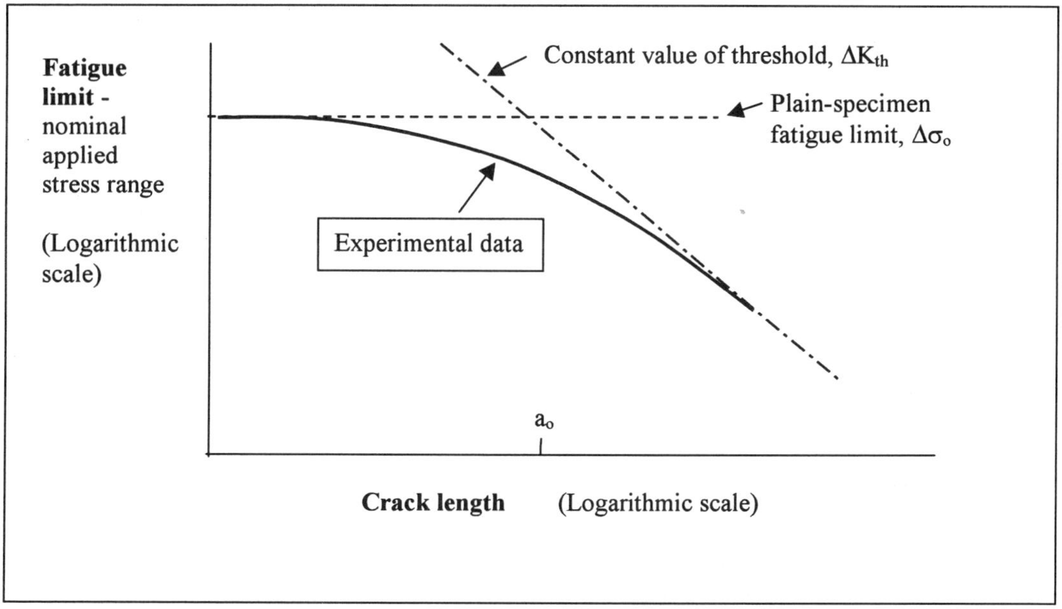

Fig.8. Illustration of the short-crack effect (Kitagawa curve)

effect potentially causes problems for the crack modelling method, since the calculation of K which it performs is correct only for long cracks, and it is difficult to modify it along the lines of equation 2, because in general the stress concentration will not be a notch of known depth. The error which is caused by not making this correction is a non-conservative error, since it leads to an overestimate of the fatigue limit. For a notch of size equal to a_o the error would be about 40%, rising to very large values for smaller lengths. The effect will be strongly material dependant because a_o varies greatly, from several millimetres in some cast irons to less than 10μm in high-strength alloys.

This error has not arisen in the two case studies described above, presumably because the features concerned were relatively large in size, giving rise to large-scale disturbances of the stress field. However, a method is needed for assessing components in order to decide whether short crack effects may play a role and, if so, to estimate their magnitude.

Conservative Nature of the Prediction

It can be argued on theoretical grounds that the original theory of Smith and Miller (1) should lead to conservative predictions, i.e. to predictions of fatigue strength which are lower than the true values, in all cases except the extreme ones: plain specimens and specimens containing long, sharp cracks, for which the predictions should be exact. The reason for this is that the approach uses two ways of calculating fatigue strength which are both inherently conservative: the use of the elastic stress at the hot spot is conservative because it takes no account of plasticity, and the assumption that the notch is crack-like is conservative because the notch will be less sharp than an ideal crack. In practice, Taylor and O'Donnell (2), having examined the data from 70 different notched specimens with K_t values ranging from 1.5 to 14, found prediction errors ranging from 45% conservative to 10% non-conservative. However, the great majority of the results (56 of the 70) fell within 5-20% conservative. These workers also examined the use of the ElHaddad method (equation 2) for describing the behaviour of short cracks, and found similar results: out of 16 data points the prediction error ranged from 10% non-conservative to 30% conservative, with 12 of the 14 points falling in the range 5-24% conservative. However it should be noted that the method of Murakami and Endo (10) which is based on an empirical fit to data from low and medium-strength ferritic steels, gives fatigue strengths which are somewhat lower than would be predicted by equation 2. If it is accepted that there will be an error of at least 10% associated with the experimental data and FE analysis, we can conclude that the above methodology will yield a prediction of fatigue strength which is correct or slightly conservative, typically by 10%. This is confirmed for real components by the results of the two case studies reported above. One final point of concern remains, however: as noted in the previous section, predictions which do not use a correction for short cracks/notches are in danger of being non-conservative to a variable degree which depends on the size of the feature concerned.

DISCUSSION

Previous methods for the analysis of stress concentrations, especially those with high K_t factors, have focussed on the problem of plasticity. The plastic nature of strain near the hot-spot has been emphasised, and methods such as the local-strain approach and elastic-plastic finite element analysis have been used to deal with it. The current approach assumes that a solution can be arrived at purely by using an elastic analysis. This, of course, is the underlying assumption of linear-elastic fracture mechanics; we can analyse a crack using LEFM, despite the fact that it has a plastic zone, provided that this zone is small, i.e. that the overall deformations of the specimen are elastic. In this work we have extended this idea to stress concentrations in components; it is assumed that the stress field calculated using an elastic analysis is characteristic of the state of stress of the component and is sufficient to allow us to decide whether failure will occur. Most components are designed for high-cycle fatigue at relatively low applied loads, and so fulfill the condition of overall elastic behaviour. Another fundamental feature of LEFM, apart from its treatment of the plastic zone, is that it is designed as a method of assessing stress singularities. The (elastic) stress field of a crack contains a singularity, and the mathematical problems raised by this are traditionally dealt with using the stress intensity concept. This implies that the method can also be efficient in dealing with other kinds of elastic singularities and near-singularities. In practice a geometrical stress concentrator cannot be infinitely sharp, but it is often convenient to model it as such in FEA: examples of such features would be weld toes, whose geometry is not precisely known or repeatable. These problems are normally dealt with by the introduction of an arbitrary, finite root radius for the feature, which is potentially very inaccurate as regards the hot-spot stress and strain. The present method, because it considers the entire stress field, can handle these singularities automatically. It should, therefore, be very suitable for problems such as welded joints.

As Smith and Miller showed, some notches are crack-like and should be assessed using fracture mechanics, whilst other notches are not crack-like: these can be assessed simply using the elastic stress at the hot spot in the classic stress-life method. The present paper has focussed on applications for which the stress concentrations were crack-like. In a more general case, it is a simple matter to decide whether a given feature is crack-like or not; one simply makes an assessment of fatigue strength using both methods and chooses the highest of the two values. This has been described and tested elsewhere (13).

A great advantage of this method is that it uses material data which should be readily available: the fatigue limit (of plain specimens) and the crack growth threshold. If necessary the threshold can be deduced from data on any specimens containing sharp notches, so the difficulties of standard fracture mechanics tests can be avoided.

This paper has emphasised aspects of the reliability of the assessment method, showing that it is not sensitive to mesh density or to the exact region of the stress field which is examined. A more serious potential difficulty is the crack/notch size effect. Further work is required in order to develop assessment procedures which include this effect.

CONCLUSIONS

1) The crack modelling method has been able to give accurate predictions of the fatigue strength of components of complex shape, both in laboratory testing and in service.

2) The method is not sensitive to the density of the FE mesh: mesh refinement which considerably altered the estimate of hot-spot stress had a negligible effect on predictions of fatigue strength.

3) Predictions of good accuracy can be obtained from different portions of the stress field. In particular, the method does not require accurate information about stresses very close to the hot spot.

4) The method is expected to give predictions which are slightly conservative, within the expected errors inherent in the experimental data and stress analysis. This is bourne out by examination of the literature on notch fatigue limits and by the results of the two case studies presented here.

5) Potential problems exist in respect to very small features, which would be expected to show size effects similar to those known for short cracks and notches. This problem did not arise in the case studies presented here, but some systematic method of assessment is needed which takes account of these effects.

ACKNOWLEDGEMENTS

The author is grateful to Rover, UK for continued support of this research over many years, and also to the Irish Government Science and Technology Agency, Forbairt and to Materials Ireland for the provision of scholarships.

REFERENCES

1) Smith RA and Miller KJ (1978) Predictions of fatigue regimes in notched components. *Int.J.Mech.Sci.* 20, 201-206.
2) Taylor D and O'Donnell M (1994) Notch geometry effects in fatigue: a conservative design approach. *Eng.Fail.Anal.* 1, 275-287.
3) Taylor D, Ciepalowicz AJ, Rogers P and Devlukia J (1997) Prediction of fatigue failure in a crankshaft using the technique of crack modelling. *Fatigue Fract.Engng Mater.Struct.* 20, 13-21.
4) Taylor D (1996) Crack modelling: a technique for the fatigue design of components. *Engng Fail. Anal.* 3, 129-136.
5) Taylor D (1997) Crack modelling: a novel technique for the prediction of fatigue failure in the presence of stress concentrations. *Computational mechanics* 20, 176-180.
6) Taylor D, Wang G, Devlukia J, Ciepalowicz A and Zhou W (1997) The analysis of stress concentrations in components: a modified fracture-mechanics approach. In *Modern Practice in Stress and Vibration Analysis,* Ed.Gilchrist, publ. Balkema, Rotterdam (Netherlands).
7) Miller KJ and de los Rios ER (Eds) (1986) *The Behaviour of Short Fatigue Cracks* EGF1, Mechanical Engineering Publications, London 1986.

8) Miller KJ and de los Rios ER (Eds) (1992) *Short Fatigue Cracks* ESIS 13, Mechanical Engineering Publications, London 1986.

9) ElHaddad MH, Dowling NF, Topper TH and Smith KN (1980) J integral applications for short fatigue cracks at notches. *Int.J.Fract.* 16, 15-24.

10) Murakami Y and Endo M (1983) Quantitative evaluation of fatigue strength of metals containing various small defects or cracks. *Engng Fract. Mech.* 17, 1-15

11) Peterson RE (1959). In *Metal Fatigue* Sines and Waisman (Eds), Publ. McGraw Hill, New York.

12) Mitchell MR. (1979) SAE/SP-79/448, publ. Society of Automotive Engineers, USA.

13) Taylor D and Lawless S (1996) Prediction of fatigue behaviour in stress concentrators of arbitrary geometry. *Engng Fract. Mech.* 53, 929-939.

FATIGUE RESISTANCE AND REPAIRS OF RIVETED BRIDGE MEMBERS

A. BASSETTI
P. LIECHTI
A. NUSSBAUMER

Institute of Steel Construction (ICOM),
Swiss Federal Institute of Technology, Lausanne (EPFL)

ABSTRACT

In 1993, the dismantling of a ninety-one-year-old railway bridge presented an opportunity to perform tests on four of its cross girders. The goal of the tests was to determine the fatigue resistance of riveted mild steel details and the effectiveness of different repair techniques. On the basis of these fatigue tests, as well as earlier tests performed at ICOM, it was found that highly prestressed riveted details subjected to a small bearing force can be considered a detail category ECCS 80. However, when the level of prestress in the riveted detail is unknown and the detail is subjected to a large bearing force, it is more appropriate to use a detail category ECCS 71.

Apart from a few exceptions, it was shown that the replacement of rivets with prestressed bolts can prevent further crack propagation of small cracks. However, this method acts only locally. A more efficient method to repair riveted elements is the bonding of pretensioned strips of carbon fibre reinforced plastic (CFRP) on the cracked element. The stiffness of the CFRP-strip and the application of a compressive stress at the crack tips causes the crack propagation to be reduced or, in some cases, stopped. Reasons for using the CRFP-laminates include their high fatigue resistance and high ultimate strength. The effectiveness of this technique is currently being verified by fatigue tests and crack propagation measurements on small scale specimens, and shows potential for use in practice.

KEYWORDS

Riveted details, repair, remaining fatigue life, old riveted bridges, CFRP, composite bonding.

INTRODUCTION

Many of the riveted bridges built during the railway development at the turn of this century are still in use today. Changes in rail traffic, increases in axle loads and general deterioration of the bridges require a re-evaluation in terms of strength, serviceability and remaining fatigue life. Often, the remaining fatigue life is found to be small due to overly conservative assumptions based on limited amount of information about the fatigue resistance of riveted details. In spite of increasing loads and some failures, there is a need to keep old bridges in service. Economically, it is not possible to replace all bridges when they attain a certain age. Furthermore, there may be a need to retain particular bridges as historical monuments. A better knowledge of the fatigue resistance of riveted details and the development of new techniques to repair and strengthen riveted bridge members damaged by fatigue, could help to extend the service life of a large number of bridges.

FULL SCALE FATIGUE TESTS OF RIVETED BRIDGE MEMBERS

Test Configuration and Description of Cross-Girders

Girders were taken from a riveted bridge built over the Hinterrhein River, Thusis, in 1901 (Fig. 1 and 2). The bridge was dismantled and replaced with a new construction in 1993. This presented the opportunity to remove six cross-girders and to perform fatigue test on four of the girders. The two remaining cross-girders will be reinforced with composite laminates (CFRP – *Carbon Fibre Reinforced Plastics*) and tested. This next research stage will verify the effectiveness of repair techniques using bonded carbon fibre-epoxy laminates on full scale fatigue damaged riveted elements (refer to section 3).

Fig. 1. Railway bridge over the Hinterrhein, Thusis, prior to removal.

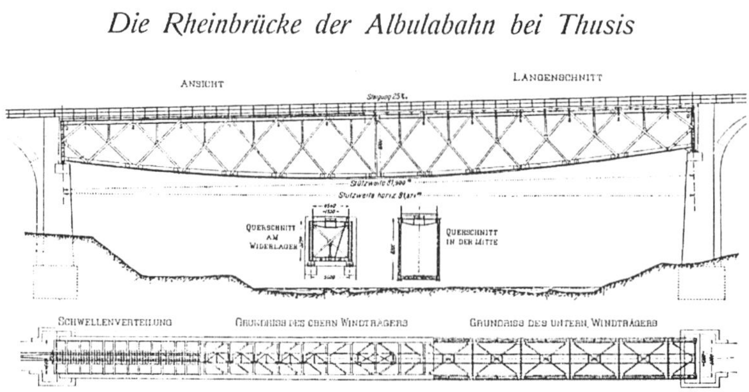

Fig. 2. Shop drawing of the steel construction built in 1901.

The dimensions of the cross-girders are indicated in Fig. 3. The top flange is composed of two angles and a riveted cover plate, whereas the bottom flange consists of only two back-to-back angles. The top and bottom flange angles were fastened at the web by a single line of rivets. All rivets were 22 mm in diameter and rivet holes were 24 mm in diameter.

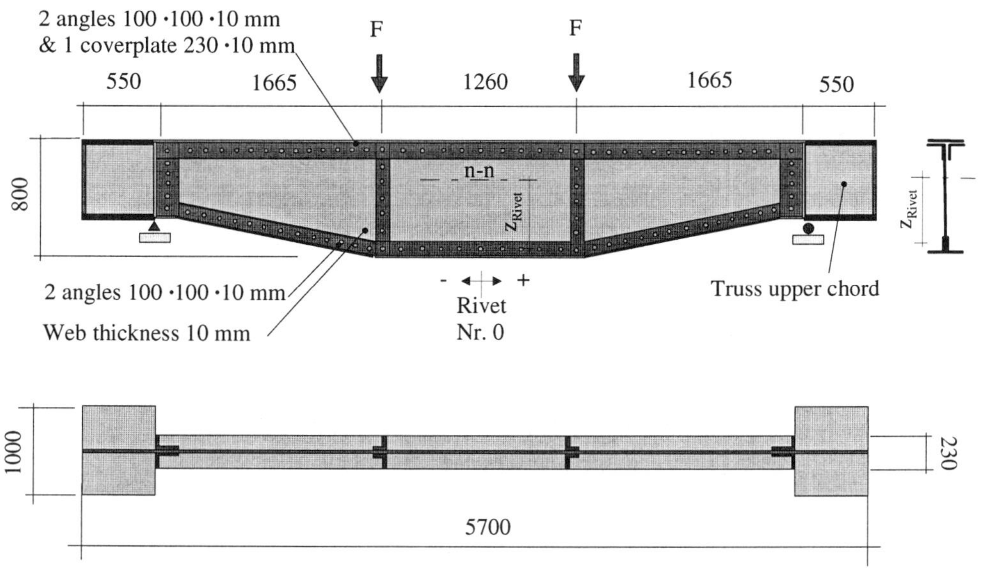

Fig. 3. Location of supports and load application on the cross-girders.

The basic material of the built-up I-shape girders consists of old mild steel. Tensile test results of six samples taken from the angles of the bottom flange are given in Table 1.

Table 1. Tensile Test Results

Property	Mean	Standard deviation	*n*
Yield strength [MPa]	302	3.8	6
Ultimate tensile strength [MPa]	412	2.3	6
Young's Modulus [GPa]	212	10.1	6

Standard Charpy tests were performed at 20°C to evaluate the fracture toughness properties. Test results from 4 samples gave a mean value for the plane strain fracture toughness, K_{IC}, of 2505 $Nmm^{3/2}$.

The single-tracked railway bridge over the Hinterrhein was in service from 1902 to 1993. An assessment of the stress range due to the effective traffic loads shows that the applied stress range probably never exceeded 46 MPa, which is lower than the damage limit of 48 MPa proposed by [1]. This means that the cumulative damage over the 91 years of service is probably insignificant.

Fatigue Testing

Four cross-girders were tested under four-point bending and were subjected to constant-amplitude load cycles at a frequency of either 2.2 or 4.5 Hertz. The location of the supports and load application points for the fatigue tests are indicated in Fig. 3. The minimum to maximum load ratio was 0.1; the applied loads are listed in Table 2.

Table 2. Test Parameters

Specimen	Stress range [(1)]	Load ratio	Loads			Frequency
	$\Delta\sigma$ [MPa]	R [-]	F_{min} [kN]	F_{max} [kN]	ΔF [kN]	f [Hz]
QT7	80	0.1	26	263	237	4.5
QT6	72	0.1	23	234	211	4.5
QT1	91	0.1	30	300	270	2.2
QT2	72	0.1	23	234	211	4.5

Note: (1) Stress range in the cross-section at the level of the rivet Nr. 0 (rivet Nr. 0: connection between web and bottom flange, at the centre line of the cross-girder, see fig. 3).

Results and Analysis of Fatigue Tests

Results of all tests are summarised in Table 3. Cracks occurred exclusively at the rivet holes. After the detection of the first cracks, the cross-girders QT1, QT2 and QT6 were repaired by replacing the rivet at the cracked hole with a pretensioned high-strength bolt. Cross-girder QT7 was repaired by placing a cover plate on the crack. After repairs, the fatigue tests were continued until further cracks at other rivets holes were detected. The rivet number in Table 3 gives an indication of fatigue cracking in the girder. Rivet Nr. 0 is located at the center line of the beam (Fig. 3). Rivets placed to the right of the center line are numbered in increasing order from 1 to 20 while rivets placed to the left side are numbered in decreasing order from –1 to –20.

The bottom flange of the built-up I-shape section consists of to back-to-back angles only. In order to compare the fatigue test results with earlier tests performed on girders with a more common detail—a bottom flange formed by two angles and a riveted cover plate—stresses were calculated at the rivets, that is, 55 mm above the bottom edge of the girders. This should allow a better comparison between fatigue resistance of rivets placed at the web and the fatigue resistance of rivets located at the cover plates.

In most cases, cracking occurred at the rivet holes situated in the constant moment region of the cross-girders. All the test were performed with a constant load range. Differences in the stress ranges reported in Table 3 are due to the different location of the rivet holes. Clearly, rivets located outside the constant moment region were subjected to lower stress ranges.

Tests results of the four riveted cross-girders are illustrated in Fig. 4. All points lie above the line for detail category ECCS 80 and confirm the results of previous ICOM test series [2] performed on riveted built-up plate girders, riveted lattice girders and rolled girders with riveted cover plates (Fig. 5). Most of those tests were performed on highly prestressed riveted details subjected to small bearing forces. The assumption of prestressing was verified for the cross-girders by the measurement of clamping forces in two rivets. The results show a stress values of 140 and 160 Mpa, which confirm the high prestress level of the rivets. Results of other studies of full scale riveted elements [3,4,5] show that when the level of prestress in a riveted detail is unknown or the riveted detail is subjected to a large bearing force, it is more appropriate to use a detail category ECCS 71.

Table 3. Fatigue Test Results

Specimen	Crack Nr.	N	Δσ [MPa]	Rivet Nr.	Cracked elements[(1)]
QT 7	1	5'817'500	80	-1	A1/A2/W
	2	5'883'900	80	4	A1/A2/W
	3	5'883'900	80	1	A1/A2/W
QT 6	1	5'694'000	72	4	A1/W
	2	12'117'100	71	6	A1/A2/W
QT 2	1	3'800'000	71	6	A1
QT 1	1	2'905'700	91	-1	A1/W
	2	3'319'100	86	-7	A2/W
	3	3'505'000	91	-4	A2
	4	4'100'500	91	4	A1
	5	4'927'700	91	-6	A2
	6	5'140'900	81	-8	A2
	7	5'788'000	91	-2	A1/A2

Note : (1) Cracked elements : A1, A2 - bottom flange angles; W - web

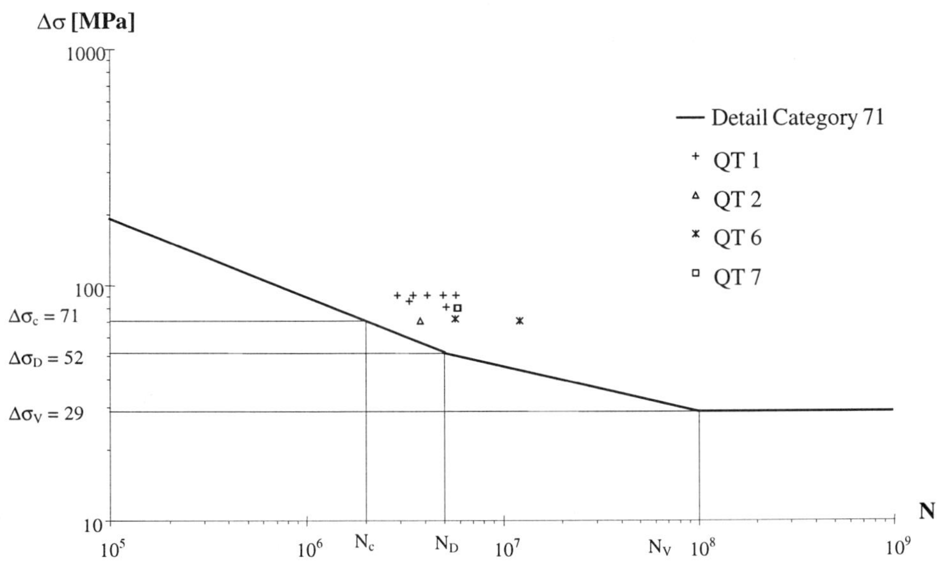

Fig. 4. Fatigue test results of four riveted cross-girders. Stress range $\Delta\sigma$ is measured in the net-section.

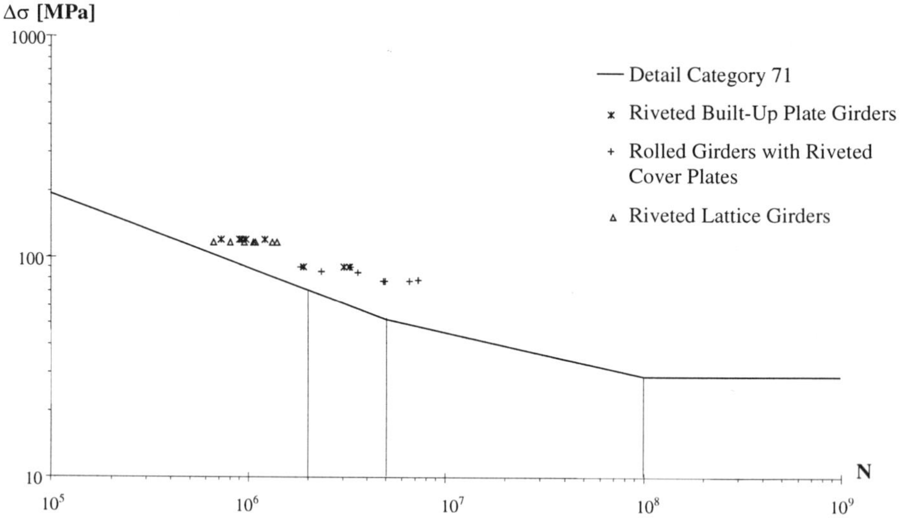

Fig. 5. Results of earlier ICOM test series of full scale riveted elements [2].

CONVENTIONAL REPAIR TECHNIQUES

Available Repair Techniques

Several techniques are available to repair bridge members damaged by fatigue:

Hole drilled at crack front. This intervention modifies the geometry and the stress distribution at the crack front, and reduces the stress intensity factor range, ΔK. This methods is quite effective in the case of distortion-induced fatigue cracking (for example by stiffeners), because the hole drilled at the crack front reduces the stiffness of the cracked member and thereby reduces the effect of secondary-moments. However, the drilled hole technique is not suitable for highly stressed sections such as truss elements, where drilling a hole will further reduce the net-section and consequently reduce the ultimate resistance. This method is also not suitable for large fatigue cracks, where, after a small period of reinitiation, the hole may form a big notch that will continue the crack propagation in the steel element.

Cover plates («patch»). This method involves placing cover plates over the crack in order to increase the section area and consequently reduce the stress range $\Delta\sigma$. However, to attach the cover plates with pretensioned high-strength bolts, it is necessary to drill holes in the cracked plate. This method is, therefore, not suitable for highly stressed sections due to the reduction of the net-section. For riveted bridge members it must also be considered that cover plates can be placed only on flat surfaces free from rivets. This measure is, therefore, not adapted for sections with a high density of rivets such as truss nodes or flanges of built-up I-sections.

Replacement of rivets with high-strength bolts. High-strength bolts can attain higher clamping forces than rivets. Higher clamping forces influence the fatigue resistance in two ways: friction forces between the joined elements are increased, thus reducing bearing forces and stress intensity ranges at the periphery of rivet holes; a triaxial stress state around the hole due to the lateral compression of the connected plates is generated. These radial and circumferential compression stresses increase the fatigue resistance of the riveted details considerably. Furthermore, by drilling the rivet holes the microcracks and defaults located around the rivet hole are eliminated. Clearly, this technique can be combined with the afore-mentioned methods, for example by drilling a hole at the crack front and filling the hole with a pretensioned high-strength bolt to reduce the possibility of crack reinitiation from the hole.

Cold expansion. This method was developed in order to increase the fatigue resistance of rails joined by bolts [6]. It is performed by cold working the hole with a mandrel and thereby increasing the hole diameter. The plastic deformations generated with this method create residual compressive stresses in the periphery of the hole, which increase the fatigue resistance of the detail. As with the afore-mentioned methods, this technique can be used to improve the initial fatigue resistance of rivet holes, and, in combination with other methods, can also be applied to stop fatigue cracks already present in the structure.

Gas tungsten arc remelting (TIG). Small cracks and imperfections with lengths smaller than 3 mm can be eliminated by remelting the metal. This methods was developed for welded constructions and is not suitable for steels of old riveted bridge members like wrought-iron and old mild steel.

Gouging of crack with air-arc, filling gouged area with weld metal. This crack control measure should be accompanied by a method that reduces the fatigue stress range in the area surrounding the crack. It is also not suitable for non-weldable steels of old riveted bridges like wrought-iron and old mild steel.

All of the above repair strategies share the following characteristics:

- Each method acts locally, on only one rivet or on only one crack.
- For the application of these repair techniques, it is necessary to detect the fatigue crack or to locate the rivet where the fatigue crack will probably propagate.

The analysis of the conventional repairs techniques has resulted in the definition of three main strategies in the repair of fatigue cracked bridge members:

1. Reduce the stress range $\Delta\sigma$ (for example by adding cover plates).
2. Reduce the stress intensity factor range ΔK (for example by changing the geometry at the crack front).
3. Apply a field of residual stresses in compression (for example by cold working the rivet hole).

Effectiveness of several conventional repair techniques

The fatigue tests performed on the four riveted cross-girders presented the opportunity to study the effectiveness of several repair methods. In order to prolong the fatigue tests, repairs were effectuated after detection of the first cracks. Cross-girder QT 7 was repaired by placing a cover plate on the crack. Other methods included replacing the rivets with high-strength bolts or drilling a hole at the crack front (as used on cross-girders QT 1, QT 2 and QT 6). The crack lengths at the time of the repair and related observations are summarised in Table 4. The test results show that in the case of small cracks, further propagation can be reduced or stopped by replacing the rivets with prestressed high-strength bolts. However, with large crack lengths, this measure is ineffective. This observation emphasises the importance of crack detection, in fact, conventional repair techniques are only effective with small cracks and can be applied only if cracks are detected in an early phase of the propagation.

A new repair technique that applies simultaneously the above three repair strategies, is currently being studied at ICOM. The following section describes this new technique that involves the application of CFRP-laminates.

Table 4. Repairs of Cracked Rivet Holes

Specimen	Rivet	N	Repair method	Cracks in flange elements[1] [mm] A1 a_u	A1 a_d	W a_u	W a_d	A2 a_u	A2 a_d
QT 7	-1	5'817'000	Cover plate	∞	∞	97	∞	∞	113
		5'918'000		∞	∞	97	∞	∞	121
QT 6	4	6'377'000	Rivet replaced with HS bolt	41					
		17'185'000		41					
		18'776'000		∞		27			
QT 6	6	12'177'000	Rivet replaced with HS bolt		22				29
		16'644'000	Hole drilled, ø 12		22				38
		17'185'000	Hole drilled, ø 22 + HS bolt		22				70
		18'776'000			∞				70
QT 2	6	6'217'000	Rivet replaced with HS bolt	38					
		7'351'000		∞					
		8'889'000		∞					
QT 2	-4	7'391'000	Rivet replaced with HS bolt + Hole drilled, ø 22 + HS bolt			86			
		8'889'000				86			
QT 1	-1	2'906'000	Rivet replaced with HS bolt			27			
		6'144'000				27			
QT 1	-7	3'319'000	Rivet replaced with HS bolt					25	
		4'010'000						35	
		4'101'000						∞	
		5'027'000				57		∞	
		5'788'000				99		∞	

Note : (1) Cracked elements : A1, A2 - bottom flange angles; W - web (Fig. 6); symbol ∞ indicates that the element was completely cracked.

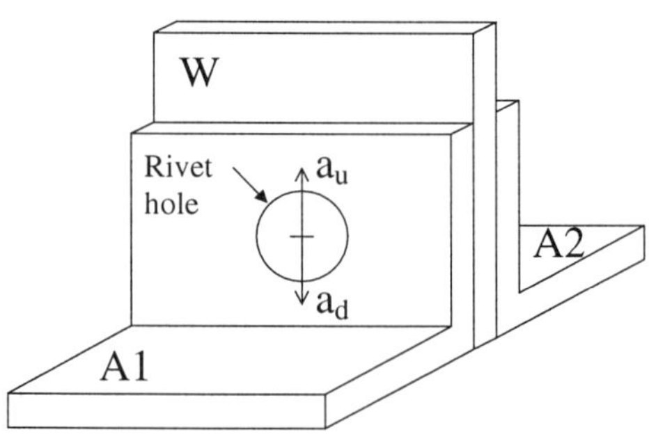

Fig. 6. Designation of bottom flange elements and directions of crack propagation.

NEW REPAIR TECHNIQUE: APPLICATION OF CARBON-FIBRE/EPOXY COMPOSITES

Principles

Carbon fibre/epoxy composites exhibit the following characteristics:

- high stiffness (Yong's modulus: $E = 155 \div 300$ GPa),
- low density (1.6 t/m^3),
- high fatigue resistance,
- high tensile strength ($f_{tk} = 1'400 \div 2'400$ MPa),
- good corrosion resistance,
- pronounced anisotropy.

Due to their qualities, composite materials have been used for many years in the automotive and aeronautical industry in order to produce structural elements subjected to extreme actions (high fatigue loads, high temperature ranges, exposure to aggressive agents). More specifically, these technical realisations show that it is possible to produce high strength assemblies using composite materials bonded to metallic elements. Fatigue test performed on aluminium [7] showed that a CFRP-patch bonded over fatigue cracks can considerably increase the remaining fatigue life of the structure. With aluminium, the effectiveness of this method is aided by the high ratio of composite stiffness ($E = 155 \div 300$ GPa) to aluminium stiffness ($E = 70$ GPa). This ratio is clearly less favourable when composite laminates are bonded to steel elements. Therefore, for application of CFRP-laminates to riveted bridge members in order to stop fatigue cracking, one of the two following strategies must be adopted:

1. Reduce the stress range $\Delta\sigma$ by increasing the section of the composite strip in order to obtain a more favourable ratio between composite strip stiffness and cracked steel element stiffness.
2. Alter the stress field at the crack front by applying thin pretensioned composite strips.

The application of pretensioned CFRP-laminates presents the following advantages:

- Due to the high tensile strength of composite materials (two times higher than normal steel used for post-tensioning tendons), it is possible to apply large pretensioning forces even in small sections. This makes it possible to carry out repairs on elements with a high density of rivets such as truss nodes or flanges of built-up I-sections.
- The application of thin strips limits the shear stress range $\Delta\tau$ between the composite and the steel produced from fatigue loading.

Compared with the conventional repair techniques described in the previous section, the application of CFRP-laminates presents the following advantages:

1. The application of CFRP-laminates does not require elements to be welded to the riveted members or holes to be drilled in the cracked plates. Hole drilling results in a reduction of the net-section.
2. Bonding CFRP-laminates to damaged members will simultaneously stop fatigue cracking and increase the ultimate resistance.
3. The applied CFRP-strip acts along its entire length thereby increasing the fatigue resistance of all covered rivets.
4. CFRP-laminates can be applied as preventive measure on bridge members that have reach the end of their theoretical remaining fatigue life, but where fatigue cracks have not yet been detected.
5. The application of CFRP-laminates does not require exact knowledge of the crack location in order to be effective.

Fatigue Tests on Small Scale Specimens

In order to verify the effectiveness of CFRP-strips bonded to riveted bridge members to stop fatigue cracking, preliminary fatigue tests were performed on small scale specimens (Fig 7).

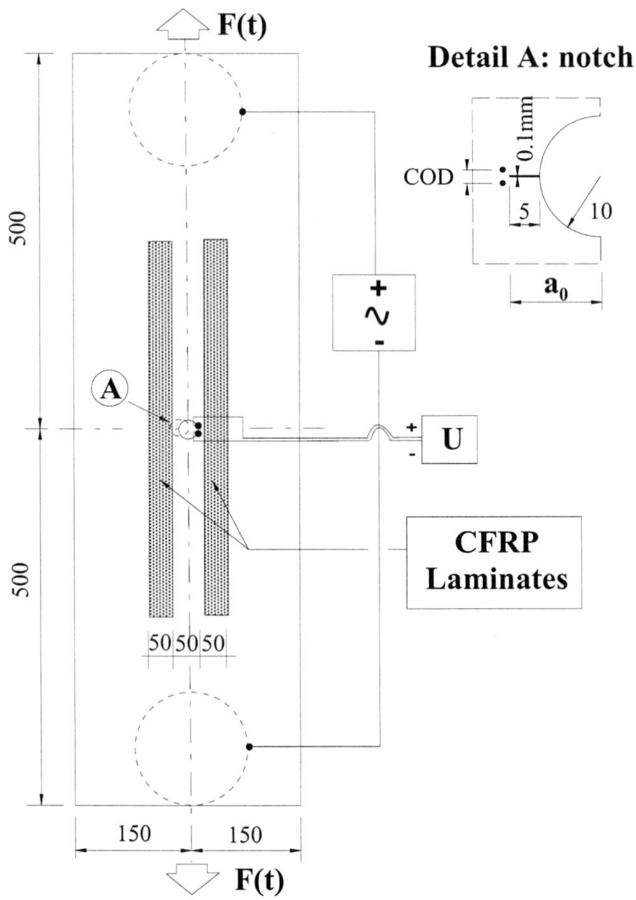

Fig. 7. Small scale specimen (plate thickness: 10mm)

The aim of this preliminary test was to study the effect of the following parameters on the CFRP technique:

- prestress-level of the CFRP-strips,
- Young's modulus of CFRP-strip,
- Young's modulus of adhesive,
- crack length at the time of repair.

Crack length and crack growth rate measurements were carried out using the electric potential drop method. COD measurements were performed in order to study the influence of the prestress level of the CFRP-strips on crack closure and to determine the effective stress range.

The experimental program and test parameters are summarised in Table 5. The preliminary results from fatigue tests performed on specimens reinforced with non-pretensioned CFRP-strips are shown in Fig. 8.

Table 5. Experimental Program

Specimen	Nr.	Parameter	Prestress [MPa]	Young's Modulus [GPa]	Strip thickness [mm]	Adhesive thickness [mm]	Strip distance [mm]	σ_{max} [MPa]	R [-]
A	3	- *(steel only)*	-	-	-	-	-	143	0.4
B	3	- *reference*	1'000	155	1.2	0.02	50	143	0.4
C	3	Prestress level	0	155	1.2	0.02	50	143	0.4
D	3	Young's Modulus	1'000	300	1.4	0.02	50	143	0.4
E	3	Strip thickness	1'000	155	1.4	0.02	50	143	0.4
F	3	Initial ΔK	1'000	155	1.2	0.02	70	143	0.4
G	3	Adhesive	1'000	155	1.2	1.0	50	143	0.4

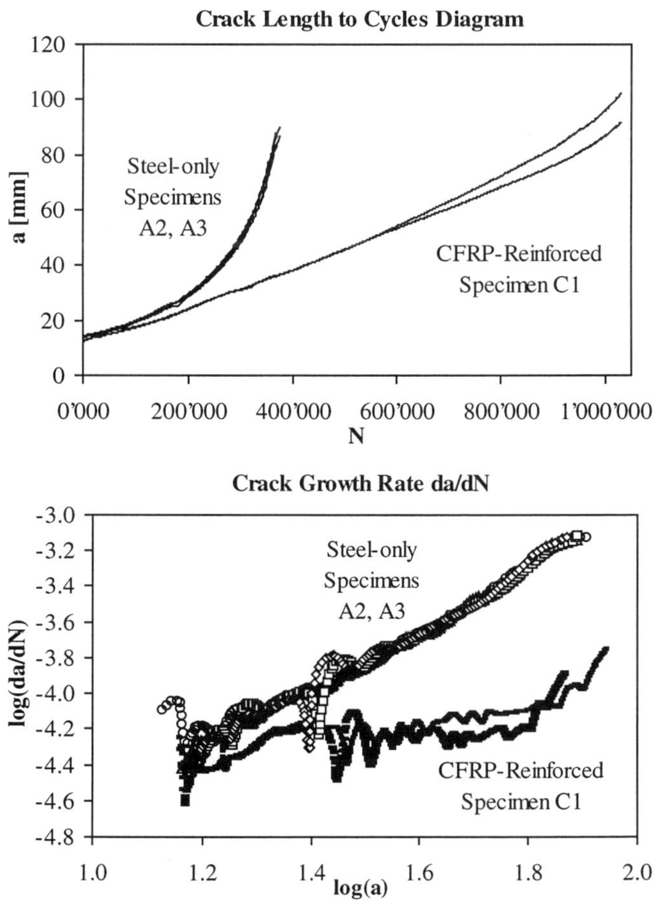

Fig. 8 Fatigue tests results of small scale specimens reinforced with non-pretensioned CFRP-laminates

Tests performed with non-pretensioned strips show that the application of low-modulus, 1.2 mm thin CFRP-laminates already significantly reduces the crack propagation rate and increases the remaining fatigue life of the specimen by a factor three. Furthermore, results of numerical simulations have shown that the reinforcement of steel members with pretensioned CFRP-strips can potentially be very effective in the domain of fatigue strengthening and repair.

ACKNOWLEDGMENTS

The authors would like to thank the Swiss National Science Foundation for sponsoring research at ICOM in the area of fatigue. Thanks are also expressed to the *«Rhätischen Bahn»* railway for supporting fatigue testing of riveted bridge members. Finally, Sika AG – Zurich is acknowledged for supplying composite materials.

REFERENCES

1. Kunz, P., Hirt, M.A., *Grundlagen und Annahmen für den Nachweis der Ermüdungssicherheit in den Tragwerksnormen des SIA*, Dokumentation D 076, Schweizerischer Ingenieur- und Architekten-Verein SIA, Zürich, 1991.
2. Brühwiler, E., Smith, I.F.C., Hirt, M.A., Fatigue and Fracture of Riveted Bridge Members, *J. of Structural Engineering*, ASCE, 1990, vol. 116, No. 1, pp. 198-214.
3. Adamson, D.E., Kulak, G.L., *Fatigue Tests of Riveted Bridge Girders*, University of Alberta, Structural Engineering Report No. 210,1995.
4. DiBattista, J.D., Kulak, G.L., *Fatigue of riveted tension members*, University of Alberta, Structural Engineering Report No. 211,1995.
5. Fisher, J.W. Yen, B.T., Wang, D., Mann, J.E., *Fatigue and Fracture Evaluation for Rating Riveted Bridges,* National Cooperative Highway Research Program Report 302, National Research Council, Washington, D.C., 1987.
6. Cannon, D.F., Sinclair, J., Sharpe, K.A., Improving the Fatigue Performance of Bolt Holes in Railway Rails by Cold Expansion, *Conference Proceedings, Fatigue Life Analysis and Prediction*, ASM, 1986.
7. Weiler, W., *CFK-Patches auf dünnwandigen, schwingbeanspruchten, metallischen Strukturen: Verhalten , Bemessung, Modellierung*, EMPA, Abteilung Ermüdung/Betriebsfestigkeit, Dübendorf, 1994.
8. Fisher, J.W., Yen, B.T., Wang, D., Fatigue Strength of Riveted Bridge Members, *J. of Structural Engineering*, ASCE, 1990, vol. 116 , No. 11, pp. 2968-2981.
9. Kunz, P., *Probabilistisches Verfahren zur Beurteilung der Ermüdungssicherheit bestehender Brücken aus Stahl*, Thèse No. 1023, EPFL, Lausanne, 1992.
10. Shulley, S., *Application of Composites to the Rehabilitation of Steel Infrastructure*, CCM Report 94-19, University of Delaware, Center for Composite Materials, Newwark, Delaware,1994.
11. Ammar, N., *Rehabilitation of Steel Bridge Girders with Graphite Pultrusion*, CCM Report 96-26, University of Delaware, Center for Composite Materials, Newwark, Delaware, 1996.
12. Sen, R., Liby, L., Spillet, K., Mullins, G., Strengthening Steel Composite Bridge members using CFRP Laminates, *Conference Proceedings, Non-metallic (FRP) Reinforcement for Concrete Structures*, RILEM, 1995.

THE SIMILITUDE OF FATIGUE DAMAGE PRINCIPLE: APPLICATION IN S-N CURVES-BASED FATIGUE DESIGN

S. V. PETINOV
Marine Technical University & Mechanical Engineering Research Institute of Russian Academy of Sciences, St.Petersburg, Russia

H. S. REEMSNYDER
Homer Research Laboratory, Bethlehem Steel Corp., Bethlehem, PA, USA

A. K. THAYAMBALLI
Research & Development, American Bureau of Shipping, New York, NY, USA

ABSTRACT

The current procedures of fatigue design of ship structural components via application of S-N diagrams based on the test results of typified welded joints physically are not complete and may need in adjustment. The main base for discrepancies of actual and S-N-modeled fatigue behavior of structural details is formed by the fairly approximate following the principles of identity of material fatigue damage in fatigue assessment of structural details.

The identity of fatigue damage in both, hull details and specimens may be limited by the initial phase of fatigue process, mostly within the stress concentration zone, specifically since the local conditions are assessed in fatigue analysis and in fatigue design of ship hull structures.

To retain the current format of fatigue design a procedure of reduction of design S-N curves is developed considering the «informative» part of fatigue process, completed with the adjustment which accounts for the possible mismatch in conditions for fatigue damage in laboratory specimens and structural details. The method is based on application of the local strain approach to fatigue assessment of both, the specimen and the detail. The adjustment suggests estimation of the ratio of fatigue lives of these and, respectively, correction of the reduced appropriate S-N diagram. The loading history of the detail comprises the feasible modes of loading and the service conditions, while the loading history of a specimen is related to the principal loading mode in a particular structure. A possible difference in fatigue properties of base-line specimen and structure materials may be considered.

KEYWORDS

S-N diagrams, fatigue design principles, fatigue of hull details, hull structures, identity of fatigue damage

INTRODUCTION

The problem of similitude and identity of fatigue damage processes exists along with the long-established and current methodology of fatigue studies and interpretation of their results in fatigue analysis and fatigue life prediction of structural details.

Specific to the methodology is the investigation of fatigue process in a particular material by testing of relatively small test pieces. The conviction that the fatigue reveals a fairly complicated process controlled by many factors the influence of which is difficult to trace in typical for structural detail conditions (e.g., [1]), results in a schematization of test conditions, typical of which is choosing the most influential factors to determine their effects on the damage. The schematization concerns the design of test pieces with respect to the actual structural detail, the choice of loading mode for laboratory tests, schematization of actual particular loading conditions, mostly by choosing the cyclic loading affected by specific limitations of the testing equipment, etc.

The means and the range of considering the most influential factors in fatigue resistance of structural details, the means of qualitative and quantitative analysis of their effects on fatigue behavior of laboratory specimens and of interpretation of fatigue properties of a structure constitute a particular structural fatigue model.

Since fatigue initiation and progress is a specifically localized process, the degree of detalization of the local conditions and the means of account for their influence would result in efficiency of a particular model and, in a more or less distinct uncertainties in evaluation of fatigue parameters.

In modeling structural fatigue one of the critical problems is to what extent the principles of similitude and identity of fatigue damage are implemented in relationship: *fatigue damage in a laboratory specimen - fatigue damage in a structural detail.*

THE BASIC PRINCIPLES

In fatigue the mechanics of material damage and failure is presented in a phenomenological format and along with definition of parameters responsible for the damage according to the present state-of-art, the principle of similitude should be substituted by the principle *of identity of fatigue damage.*

The principle implies that the fatigue damage in structural detail at a location where the damage likely to occur should be identical to the damage in a laboratory specimen. It means that the material structure composition and condition, surface conditions, the load traces, the presence of static loading modes, etc., i.e. conditions for development of fatigue processes and *the state of fatigue damage must be identical in both.*

In choosing the appropriate format for description and modeling the local fatigue damage an important feature of alternating stress field should be considered: in structural discontinuities material is subjected to the loading conditions which may be approximated as the loading under the displacement range control. This is due to elastic behavior of surrounding material of structure acting as a rigid testing machine, due to the redundancy of the most structures, the compatibility of which is not affected in the initial stages of the consecutive crack growth.

The S-N diagrams accepted in fatigue analysis and design are the result of fatigue tests in which the failure criteria was «cycles to separation into two pieces» [2]. By this defined, the fatigue life of a specimen includes the crack initiation and early propagation when the damage is influenced by geometry, residual welding stress and particular properties of material at origination site, and the proceeding crack propagation which is controlled mostly by the loading conditions and increasing compatibility of a specimen. These observations, otherwise common enough, may reveal a serious mismatch between the damage conditions in a structural detail and in related base-line specimen.

Therefore it may be concluded that the identity of fatigue processes in both cases is restricted to the initial phase of the process including fatigue crack initiation and early propagation, mostly within the stress concentration zone.

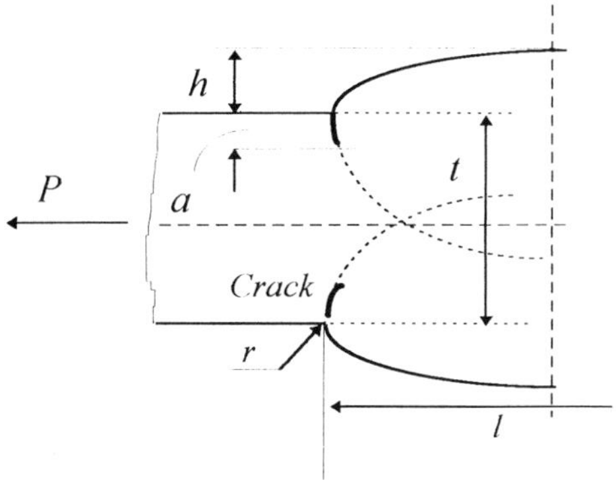

Fig.1. Geometric parameters of butt-welded joint and the mode of loading

To illustrate this an example might be given. Assume a butt-welded joint typical of aluminium panel structures of hydrofoil crafts subjected to repetitive tensile in-plane loading perpendicular to the weldment, Fig.1. The weld toe geometry is assumed as smooth, with the relative toe radii representative for the experimentally estimated range, $r/t = 0.03$ and 0.06. The other weld shape proportions are taken as $l/t=3.0$; $h/t=0.5$, typical of welded joints in Al-Mg panels. Assume also that the fatigue process is governed by the crack propagation. In an idealized case fatigue cracks originate simultaneously at the weld toe from both surfaces of the plate. The crack growth may be approximately analyzed using Paris-Erdogan equation and corrections for the stress concentration influence; the decrease of stiffness is not considered assuming the strain-controlled mode of loading (due to redundancy of the whole structure).

The FEA resulted in stress concentration factors, $K_t = 2.13$ and 2.60, at the above weld toe radii, respectively. The stress intensity factors were calculated using the approximate relationship:

$$\Delta K = \Delta\sigma_n(\pi a)^{1/2} M_{sc}(K_t, a) \tag{1}$$

where σ_n is the nominal stress amplitude, $M_{sc}(K_t,a)$ is the correction for the stress concentration influence. To find this, the dimensionless principal stress distribution along the expected crack path was approximated according Testin et al [3]:

$$M_{sc}(K_t,a) = 1 + (K_t - 1)\exp(-35(K_t - 1)2a/t) \tag{2}$$

In stress intensity calculation the plasticity correction was neglected on assumption that at relatively low stress concentration and moderated stress amplitudes, $0.3 \le \sigma_n/\sigma_{0.2} \le 0.6$ ($\sigma_{0.2} = 180$ MPa) the crack tip plastic zone is small. The number of load cycles the crack would need to reach a certain size is found by integration of the Paris-Erdogan equation:

$$N(a) = (1/C)\int_{a_0}^{a} (\Delta K)^{-m} da \tag{3}$$

the parameters of which were obtained from results of fatigue testing of CT specimens: $C = 6.09 \times 10^{-11}$, $m = 2.7$. The upper limit in the integral (3) is a variable and its maximum value is limited in the range up to $a_{max} = t/3$. The initial crack size may be found from the Eqn (1) on assumption that that the initial crack size was equal to the size of effective plastic zone at the notch root. The stress amplitude at the boundary of effective plastic zone ($a = a_0$) was estimated approximately as an average of conditional proportionality stress in the generalized cyclic curve of aluminium alloy AlMg-61, $\Delta\sigma'_p$, and of fatigue limit stress related to N = 10^7 cycles; the latter according to [4] is : σ_{-1} (10^7) = 50 MPa. Consequently, $\sigma(a_0) = 0.5(\Delta\sigma'_p/2 + \sigma_{-1}(10^7)) = 110\,\text{MPa}$, and thence the initial crack size:

$$a_0 = -\left(t/\left(70(K_t - 1)\right)\ln\left(M_{sc}(a_0) - 1\right)/(K_t - 1)\right) \tag{4}$$

where $M_{sc}(a_0) = \sigma(a_0)/\sigma_n$. Another estimate of the magnitude of a_0 may be found from [5]: $a_0 = 0.0198(ta_m)^{1/2}/\alpha$. The parameter a_m was found from Peterson's formula $a_m = 0.38(350/\sigma_b)^{1.16}$; the parameter α depends on the type, geometry of weld and loading mode. According to [3] α = 0.23 for the axial loading mode of butt welds.

The fatigue strength reduction factor was defined as

$$K_f(a_i) = \sigma(N)/\sigma_n(N(a_i)) \tag{5}$$

where $\sigma(N)$ is the stress amplitude corresponding to material failure at N cycles, $\sigma_n(N(a_i))$ is the nominal stress amplitude. The stress amplitude $\sigma(N)$ and the related fatigue life were calculated using the local strain and the LEFM formats, respectively. In the initiation part of the process the strain-life criterion, Neuber's formula and the generalized cyclic curve were used to convert the strain-life criterion into the nominal stress-life relationship.

The results of calculation are given in Fig.2. It is seen that the crack growth is controlled by the stress concentration at a distance which is about of three times the weld toe radius, up to 0.5 - 0.7 mm.

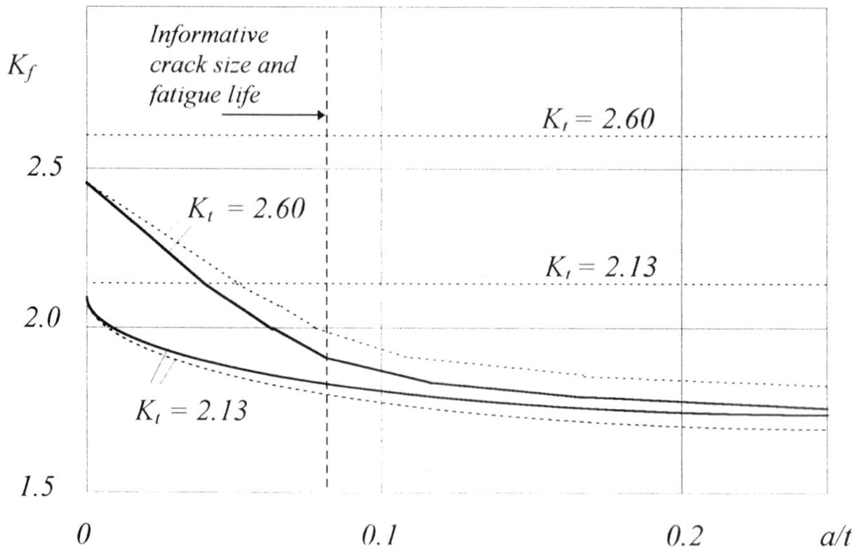

Fig.2. Fatigue strength reduction factor vs crack size (N = 10^4): Solid and dashed lines are related to different methods of evaluation of the initial crack size

Beyond this depth the crack is insignificantly influenced by stress concentration and the fatigue strength reduction factors for different weld geometries reach approximately the same magnitude, depending on the component compatibility, $K_f \cong 1.7$. It means, that in fatigue design of a structural component the «informative» crack size should not extend over the stress concentration-affected region in case the target of the design is the allowable local stress, allowable stress concentration factor and geometry of critical location.

Although this is only an illustrative example it unambiguously shows: the fatigue design needs in the fatigue data related to the early, «informative», stage of fatigue crack growth, until the crack propagates within the stress concentration zone. Similarly, many of known experimental results may be interpreted to demonstrate the importance of only the initial portion of fatigue life of base-line specimens for application in fatigue design.

One more example should be given to illustrate the importance of difference in the loading conditions of laboratory specimens and of actual hull details. Assume a butt-welded steel plate loaded in tensile mode realized as (a) a pulsating load-controlled cyclic tension and (b) zero-maximum displacement-controlled testing.
Assume that initial through-thickness crack at a weld toe already exists. It might be a surface shallow crack - for the purposes of illustration it is not important. To analyze the crack propagation a version of the local strain approach might be applied. Assume also a material element located at a distance δa ahead the crack tip; the crack length is a. The size of element should allow application of continuum terms and the damage in it should be defined corresponding to failure of the element when the crack tip extends into the element position. The crack tip strain field component, the principal strain at a distance r from the crack tip ($0 \le r \le \delta a$) is $\varepsilon(r)$. The Coffin-Tavernelli's [6] material fatigue failure criterion may be written as:

$$N(\varepsilon) = C^{1/\alpha}(\varepsilon(r) - \varepsilon_0)^{-1/\alpha} \tag{6}$$

where C and α are the material fatigue parameters and ε_0 is the elastic component of the total strain range $\varepsilon(r)$ within which the damage process is ineffective. Assume that the damage due to the inelastic cyclic straining accumulates in the every load reversal according to the current inelastic strain range. Then the damage "per load reversal" accumulated at the crack extension δa is:

$$d = \Sigma \frac{1}{N(\varepsilon)} = \int_{\varepsilon_{\min}}^{\varepsilon_{\max}} \frac{d\varepsilon}{N(\varepsilon)} = \int_{a}^{a+\delta a} (N(\varepsilon))^{-1}(d\varepsilon / dr)dr \tag{7}$$

and the damage accumulated in through the δn load reversals can be written as

$$d(\delta n) = \delta n C^{-1/\alpha} \int_{a}^{a+\delta a} (\varepsilon(a,r) - \varepsilon_0)^{1/\alpha}(d\varepsilon / dr)dr \tag{8}$$

This expression characterizes the part of damage to be accumulated at the crack extension δa. Actually, the damage summation commences from the moment when the boundary of plastic zone, r_p, reaches location of material element due to crack progress. Therefore the total damage consists of two components: one which is given by eqn (8) and another, accumulated at the crack extension over the distance $r_p - \delta a$,

$$d_p(\delta n,\delta a) = (\delta n / C^{1/\alpha}) \int_a^{a+\delta a} \varepsilon_{ef}^{1/\alpha}(d\varepsilon/dr)dr + (n_a / C^{1/\alpha}) \int_{a+\delta a}^{a+r_p} \varepsilon_{ef}^{1/\alpha}(d\varepsilon/dr)dr \qquad (9)$$

where $\varepsilon_{ef} = \varepsilon(a,r) - \varepsilon_0$ is the effective, plastic plus microplastic, strain range.

To solve the problem, a FE-model of the plate was prepared using the isoparametrical quadrilaterals and to model the crack tip field the mesh was automatically re-adjusted at the crack increments. To model adequately cyclic plasticity the flow theory was applied in the software. Material of the plate - mild steel of the 235 grade. The parameters of failure criterion (6) were obtained experimentally: C = 0.636, α = 0.654; continuously recorded cyclic curves were used to define the generalized stabilized cyclic diagram. For comparison the fatigue process may be simulated in the LEFM format within the range of feasible evaluation of stress intensity factors. The parameters of Paris-Erdogan equation used were: C = 2.5×10^{-13} , m = 2.88. The results of the crack growth estimation using the both approaches are given in Fig.3.

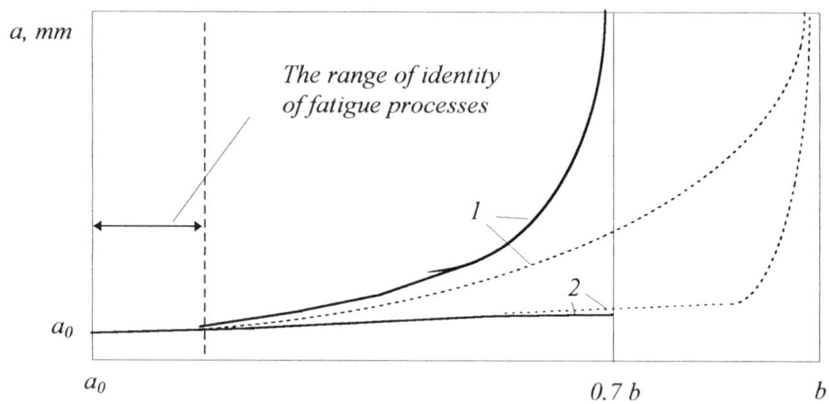

Fig.3. Fatigue crack growth curves: Solid lines - calculated using LEFM-format, dashed lines - local strain approach. (1) - the load control; (2) - displacement control. The LEFM applicability range is limited by the condition $a \leq 0.7b$

These results demonstrate the uniqueness of the crack propagation in initial phase and the difference in the crack growth behavior in the both loading conditions which develops from the moment when the increasing compatibility of the «test piece» starts to accelerate crack in the load-controlled mode of «numerical testing» the plate. Application of the both approaches results in the same crack growth laws, the local strain approach, however, provides analysis of the process until complete failure of specimen. It may be concluded from these results that the damage identity is limited by the period of crack growth when the rate of the process does not depend on the loading conditions.

Again, this is only a cause to initiate detailed discussion and analysis of conditions for splitting the crack growth behavior in typical non-continuous welded joints. However, it can be seen, that to form the appropriate data base for fatigue design *the design state of fatigue damage* must be introduced related to the influence of stress concentration and damage identity phase. With this, the rules for future fatigue testing and for editing the existing fatigue data (S-N diagrams) might be established.

The following conclusions might be drawn from the above arguments and discussion:

- fatigue testing of specimens should preferably be carried out under the cyclic displacement range control conditions
- fatigue life of specimens for the purposes of fatigue design of structural details should preferably be limited by the number of cycles corresponding to the «informative» crack size at a notch

Analysis of the early crack growth reveals the high growth rate within the mentioned phase, decreasing though, when the crack tip extends from the stress concentration zone [7], [8], etc. Therefore it is possible to assume that fatigue failure of the plastic zone at the stress concentration occurs almost simultaneously. This assumption introduces certain conservatism into estimation of the rate of process, on the one hand, and, on the another, permits to apply the results of fatigue testing of uniformly stressed specimens under strain range control as typical of fatigue behavior of material at stress concentration, as it was shown by numerous analyses [9], [10], etc.

Therefore the approach based on the strain criterion of fatigue failure and methods of evaluation the local strain at a notch in structural detail may be preferable in fatigue analysis of structures. Additionally, the effects of the cyclic strain hardening, of the load ratio due to applied loads and due to influence of residual welding stresses on the strain range in current load reversal should preferably be considered by introducing an *effective strain range*. Consequently, the fatigue identity condition for the initiation phase additionally to the above may be assumed as the following: *the effective local strain range should be identical at the crack origin in structural detail and in laboratory specimen.*

However, the local strain approach needs in further updating to consider stochastic conditions for crack initiation in welded joints, in assessment of the non-uniform microstructural plasticity related to the high-cycle portion of fatigue diagrams.

Although the means of solution of these problems are seen, first the evolution in the fatigue design procedures should be suggested. It may be realized as retention of the current methodology completed with the procedure of reduction and adjustment of the design S-N diagrams.

REDUCTION AND ADJUSTMENT OF THE DESIGN S-N DIAGRAMS

A complication may be seen in definition of the standard joint classes: the classes most important in the scope of fatigue design, incorporating the non-continuous welds are the result of lumping experimental data related to essentially different geometries [2]. To solve the problem the data sets first should be edited so that the information would reflect principally effects of the particular weldment geometry and of the loading mode. After this the S-N diagrams related to specified geometries and loading conditions should be reduced by computing off the crack growth history from total fatigue lives so that resulting life would reflect the effects of stress concentration in the edited diagrams. This might be done by applying a FEM-based combined technique in which the high-cycle range of S-N diagrams (when total mean lives exceed, approximately, $N = 10^5$ cycles) must be re-analyzed with the aid of LEFM methodology. At shorter lives and in the final phases of the high-cycle crack growth preceding the separation in the two pieces when the material plasticity becomes a principal factor at the crack tip, a version of the local strain concept in conjunction with the damage counting procedure demonstrated in the above example and suggested earlier [11], [12], may be applied solely or in combination with LEFM format [13]. An extensive numerical analysis to distinguish particular geometries and define the design diagrams meeting requirements of the damage identity principle may be carried out using the mentioned methodologies.

Such way defined, the reduced diagrams would need in further adjustment to account for the difference in the geometry, stress ratio, load combination, load history, effects of material plasticity in severe notch conditions, etc. The procedure of adjustment of the S-N diagrams and related soft-ware [14] now may be used in evaluation of appropriate S-N curves for fatigue design.

The procedure considering effects of the above factors in the form of correction is based on application of the local strain approach in its version related to the random loading case. In general, the correction should be found as the ratio of fatigue lives of a detail under the scope and that of a reference specimen.

The procedure of correction in ship technology applications, briefly, consists of the following parts. Firstly, the stress analysis should be performed for the detail under consideration and for the relative base-line specimen. In finite-element analysis the type of mesh and of the element, the element sizes at the anticipated crack origin should be identical. The due attention should be paid to evaluation of the boundary conditions for the detail, to proper modeling of the load transfer in the detail elements; all influential modes of loading must be considered. According to results of the stress analysis the stress concentration factors (corresponding to the elastic stress field) should be obtained for both, the specimen and the detail.

In general, hull details deformation results from simultaneous application of several loads, global (bending moments in vertical and horizontal planes, etc.) and local (hydrodynamic pressures, cargo inertia loads, etc.). Resulting from the same sources (passage of wave, wave-induced ship motions, etc.) these loading modes produce specific response of a structural detail, individual components of which are characterized, even in the case of regular (harmonic) loading regime by more or less developed phase lags. These phase lags depend on the frequency of excitation (respectively, on the sea state), on the ship speed, on the heading angle, etc. The mentioned factors made it necessary to consider a series of stationary loading regimes and to compose the long-term distributions of the combined local stress parameters considering for the detail and fatigue affected zone location.

The local stress variance in a stationary sea conditions is expressed as

$$D_\sigma(h) = D_{\sigma,k}\left(1+\sum_i\left(\frac{A_i K_{t,i}}{A_k K_{t,k}}\right)^2\frac{D_i}{D_k}+2\sum_i\sum_j\delta_{ij}k_{ij}\left(\frac{A_i A_j K_{t,i} K_{t,j}}{\left(A_k K_{t,k}\right)^2}\right)\left(\frac{\sigma_i\sigma_j}{D_k}\right)\right) \tag{10}$$

where h is the characteristic wave height for assumed sea state, $D_{\sigma,k}$ is the local stress variance related to the "k"-the loading mode taken as the principal, e.g. the wave-induced bending moment in the vertical plane, D_i are the wave load variances, A_k, $K_{t,k}$ are the compatibility and the stress concentration factors corresponding to the loading modes, k_{ij} are the correlation coefficients for pairs of the random wave loads, $k_{ij} = K_{ij} / \sigma_i\sigma_j$, K_{ij} is the covariance of the «i» and «j» wave load components, σ_i and σ_j are the standard deviations of the loads and δ_{ij} is the Kronecker symbol.

In evaluation of long-term distribution of local stress range for a detail and related reference specimen a common "full-probability" procedure is used, e.g., [15]. According to assumed probabilities of the expected routes, the sea states, still-water load cases, etc., the long-term distribution of the local stress should be obtained which is presented usually in the form of the two-parameter Weibull approximation.

According to (10) the variance $D_\sigma(h)$ can be presented in the form of

$$D_\sigma(h) = D_{\sigma,k}(h,\ldots)d(k_{ij},\ldots) \tag{11}$$

where $D_{\sigma,k}(h,\ldots)$ is the variance of the local stress attributed to the "k" loading mode selected as the principal one, $d(k_{ij},\ldots)$ is the correction factor which considers the influence of the secondary loading modes; the contents of it is seen in (10).

The following step consists of composing the «weighted» variances $D_{\sigma}(h)$ according to the frequencies of occurrence of the sea conditions, heading angles, etc., to form the long-term distribution of the local combined stress for the detail of interest and for the base-line specimen, in tbe latter case mere by assuming the correction factor as $d(k_{ij},...) = 1$.
At known long-term loading history the fatigue life of reference specimen is found applying the linear damage accumulation rule,

$$N^{*}_{sp} = D/\int\left(p_{sp}(\sigma)/N(\sigma)\right)d\sigma \tag{12}$$

and, correspondingly, fatigue life of the hull detail:

$$N^{*}_{det} = D/\int\left(p_{det}(\sigma)/N(\sigma)\right)d\sigma \tag{13}$$

where D is the index of fatigue damage, identical in both cases, $p(\sigma)$ is the probability density function defined from the above long-term distributions, $p(\sigma) = -dQ/d\sigma$, $N(\sigma)$ is the material strain range-fatigue life relation:

$$\Delta\varepsilon = CN^{-\alpha} + BN^{-\beta} \tag{14}$$

This form of criterion is preferred in comparison with (7) since it approximately covers the high-cycle range which is necessary in the corrections. Parameters of (14) should be taken accordingly the fatigue properties of materials used in specimen and in hull detail. The total strain range, $\Delta\varepsilon$, should be determined using the stabilized cyclic stress-strain diagrams and an appropriate method of solution of the elastic-plastic problem. Consequently, the material fatigue failure criterion (14) may relate fatigue life of material at the crack initiation site to the total elastic stress range, $\Delta\sigma$.

Once the fatigue lives of the detail and specimen are calculated, the corrected base-line fatigue curve for the detail can be obtained. The reduced standard design S-N curve (the crack propagation cycles off the stress concentration zone are deduced from the fatigue life) is approximated in the form of Basquin's equation:

$$\log N = \log C - m\log S\,, \tag{15}$$

On assumption the slope of S-N diagram, m, (whether it is the one-slope or the two-slope diagram) is not affected by correction, the corrected parameter C of equation (15) for the detail is

$$C_{cor} = C_{sp}\left(N^{*}_{det}/N^{*}_{sp}\right) \tag{16}$$

It s possible to convert correction to reduced base-line S-N curve into a corrected allowable stress range for a particular location: it follows from (15) that the corrected allowable stress range is

$$\log S = \log S_{o} - (1/m)\log k \tag{17}$$

where S_o is the allowable stress range, a reduced Fatigue Classification for welded joints value, and $\log k = \log(N^{*}_{det}) - \log(N^{*}_{sp})$.
The corrected diagram should be used in fatigue analysis according the accepted rules, however, the hot-spot stress evaluation now is unnecessary since the effects of the "global" stress concentration are considered in the procedure.

To facilitate application of the reduction and correction of S-N curves in fatigue design a PC - programs FATICA is developed.

CONCLUSION

The problem of identity of fatigue damage in a structural detail and in relative test piece is discussed. It is shown that the «informative» part of fatigue life of test specimens should be limited to the initial phase of fatigue crack growth until the crack propagates under influence of the local stress elevation conditions. The conclusion is confirmed by analysis of fatigue process in cyclic loading and in cyclic displacement conditions, the latter more typical of hull structural details.
To provide the designers with physically more advanced technology a procedure of reduction and correction of the design S-N curves is suggested The procedure of reduction converts the basic diagram into a crack initiation S-N curve related to effects of local conditions on fatigue properties of welds. The following procedure of correction of reduced S-N curves for non-continuous welded joints considers for the mismatch in loading modes, in geometry of the base-line specimens and hull details related to the same S-N class. It allows for the inelastic response of material at critical location under alternating loading and for the effects of multi-mode loading.

ACKNOWLEDGEMENTS

The present work was partly supported by the Russian Academy of Sciences, Fund 589, and American Bureau of Shipping. The views expressed in this paper are those of the authors, and not necessarily those of the institutions they are affiliated with.

REFERENCES

1. Manson, S.S. (1965). In: *Experimental Mechanics*, 7, No.5, pp.193-226
2. Reemsnyder, H.S. (1997). In: *Proceedings of the Symposium and Workshop on the Prevention of Fracture in Ship Structures*. Washington, D.C., pp.264-327
3. Testin, R.A., Yung, J.-Y., Lawrence, F.V., Jr., and Rice, R.C. (1987). *Welding Res. Suppl. to the Welding Journal*, Apr., pp.93-98 S
4. Ziganchenko, P.P., Kuzovenkov, B.P. and Tarasov, I.K. (1981). *Hydrofoil Crafts: Strength & Construction*. Sudostroenie Pubs, Leningrad (in Russian)
5. Burnside, O.H., et al. (1984). *Ship Structure Committee*, SSC-326
6. Coffin, L.F. and Tavernelli, J.F. (1962). ASME, Ser.D, 4, 533
7. Smith, R.A. (1983). In: *Short Fatigue Cracks, ASTM STP 811*, pp.264-279
8. Miller, K.J. (1993). *Materials Science and Technology* 9, pp.453-462
9. Wetzel R.M. (1968). *Journal of Materials*, JMSLA 3, No.3, pp.646-657
10. Reemsnyder, H.S. (1986). In: *Case Histories Involving Fatigue and Fracture Mechanics, ASTM STP 918*, pp.136-152
11. Ellyin, F. and Fakinlede, C.O. (1985). *Engineering Fracture Mechanics* 22, pp.697-712
12. Petinov, S.V. (1996). In: *Proceedings of the International Offshore and Polar Engineering Conference, ISOPE-96*, Los Angeles
13. Petinov, S.V. (1997). In: *Proceedings of the Second International Conference on Marine Intellectual Technologies, MARINTECH-97*, St.Petersburg. Vol.4, pp.90-94
14. Petinov, S.V. and Thayamballi, A.K. (1998). *Due to appear in the March issue, J. Ship Research*
15. Mansour, A., Lin, M., Hovem, L. and Thayamballi, A. (1993). *Ship Structure Committee*, SSC-368

PROBABILISTIC FRACTURE MECHANICS APPROACH FOR RELIABILITY ASSESSMENT OF WELDED STRUCTURES OF EARTHMOVING MACHINES

H. JAKUBCZAK
Warsaw University of Technology, Institute of Heavy Machinery Engineering, Warsaw, Poland

ABSTRACT

Monte Carlo simulations of fatigue in welded components of excavator working attachments have been performed with the fracture mechanics approach. Variability due to local weld toe angle and radius, the crack growth rate and a scaling factor of the service loading were considered. Predictions of fatigue life for 80% of probability of no-failure were made for components fabricated by two different manufacturers. The results have indicated a distinct effect of the manufacturing quality on the fatigue life of components. It was also shown, that variability in the local geometry has the greatest influence on the predicted fatigue life of components.

KEYWORDS

Welded structures, crack growth life, probabilistic approach, Monte Carlo simulation

INTRODUCTION

Load carrying structures of earthmoving machines like excavators, loaders, etc. undergo in service random loading, and are prone to fatigue failure due to stress concentration in welded joints. Appropriate fatigue life predictions of their structural components are made at different stages of the product development, to assure their necessary resistance against fatigue failure and a safe operation in the required lifetime. Various life prediction approaches can be used for that purpose depending on data available and the problem definition [11, 13].

The nominal stress and notch root approaches have been used for dimensioning of structural components of hydraulic excavators being considered and the fracture mechanics approach has been used for assessment of their residual lives [13]. A deterministic crack propagation model was adopted for that purpose and acceptable results have been obtained. However, in some cases cracks were observed in service long before the required operation period was over. Differences in manufacturing quality of welded joints were considered a possible reason for that phenomenon since the components have been fabricated by two different manufacturers. To account for uncertainties in main parameters of the life prediction method, a probabilistic fatigue life analysis was carried out for the selected component of a hydraulic excavator. The analysis based on the Monte Carlo simulation was performed, and the residual life, spent on the crack growth from an initial crack was considered.

ANALYSIS OF FATIGUE LIFE

Object description

The boom of a 16t hydraulic excavator was considered in the present analysis. It was made from a 15G2ANb low alloy steel of the yield strength R_e= 350 MPa and ultimate strength R_m = 550 MPa. Parameters of Paris equation describing the crack growth rate of the considered material are C = 1.03e-12, m = 3.89 and K_{Ic} = 52, where da/dN is measured in [m/cycle] and K in [MPa√m] [8].

All welds have been made using manual MAG welding. Measurements of the local geometry of selected welded joints on the components have been carried out using the replica method [14]. Uncracked components have been used for that purpose. Results for fillet welds of the cover plate connection are presented in Fig. 1 as probability distributions of the weld toe angle θ and radius ρ. Concentration of most of data points at smaller values of the weld toe radius indicates the worse quality of manufacturing of the component M.2, since it may result in a higher stress concentration at the weld toe and hence lower fatigue life.

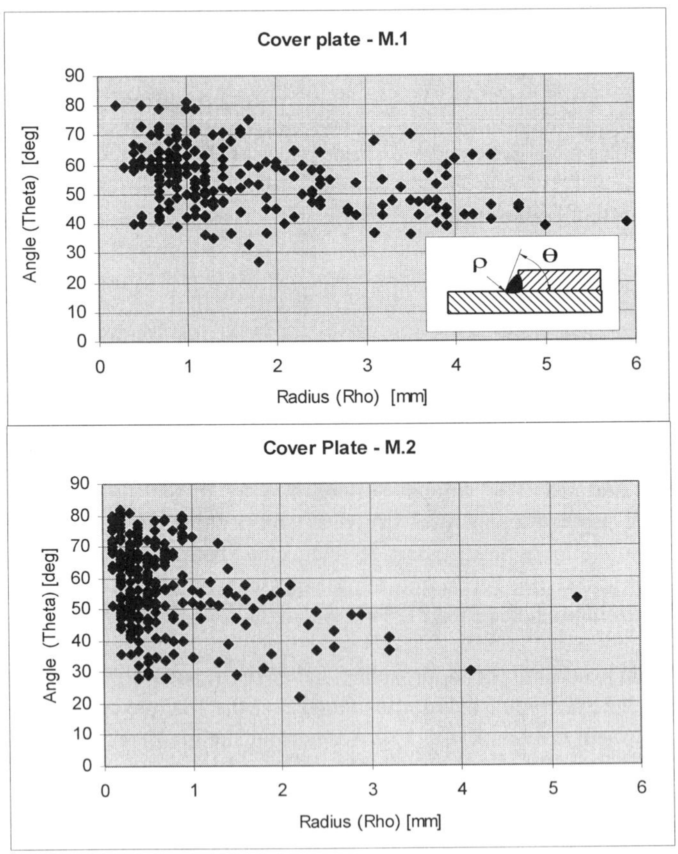

Fig. 1. Variations in the angle of inclination and weld toe radius for components made by two manufacturers.

Crack propagation model

Linear elastic fracture mechanics was used in the fatigue life analysis of the components. Lecsek at al. [9] have shown that the multiple crack growth model gives a better life prediction of welded joints than models using a single semi-elliptic surface crack or edge crack, which over-estimate or under-estimate the fatigue life giving the upper and lower bounds. Hence in the analysis of fatigue crack growth of the considered component both extreme models have been used, i.e. crack growth of a single semi-elliptic surface crack (Fig. 2a) and an edge crack was considered (Fig. 2b).

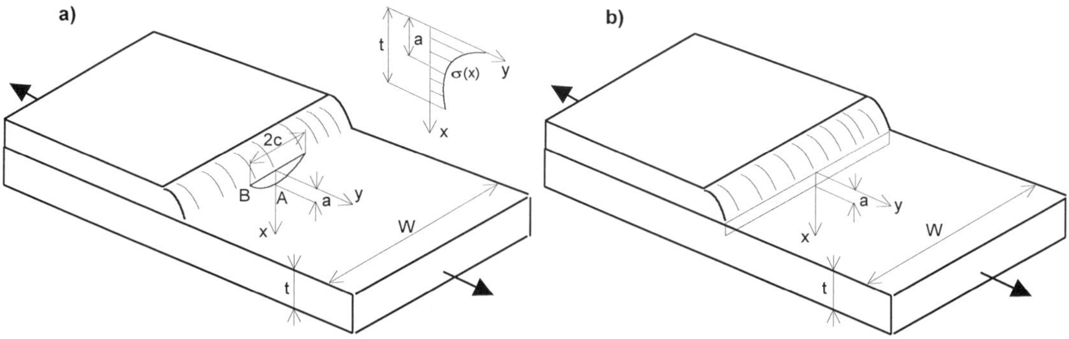

Fig. 2. Modelling the crack at the weld toe.

To calculate the stress intensity factor for a crack emanating at the wed toe one has to account for the stress distribution along the prospective crack path. The weight function method is very suitable for that purpose. The Shen and Glinka [12] weight function for semi-elliptic cracks, while the weight function for edge cracks modified by Jakubczak and Glinka [7] can be proposed here. Calculation of stress intensity factors using weight functions is very efficient. However, a problem arises if it must be repeated many times for cracks emanating at a randomly variating weld shape. This is due to lack of simplified methods of obtaining the required stress distribution through the plate thickness. The closed-form formula proposed by Niu and Glinka [10] is for bending, while those proposed by Xu at al. [17] are for symmetric notches only.

Therefore the correction function method has been used for calculating stress intensity factors in the analysis:

$$K_I = S\sqrt{\pi a} \cdot F \tag{1}$$

where F is the correction function proposed in [2]:

$$F = F_E \cdot F_S \cdot F_T \cdot F_G \tag{2}$$

The correction functions F_E, F_S, F_T and F_G account for the semi-elliptic shape of the crack, free surface, finite thickness and stress distribution respectively, whereas F_G depends on the stress concentration factor K_t.

The shape of a surface crack is modelled by a semi-ellipse and it is defined by the aspect ratio *a/c*, where *a* denotes for the crack depth and *c* for crack length at the surface. When a semi-elliptic crack placed at the weld toe grows under cyclic loading, its shape changes continuously. The surface point of the crack always undergoes the high stress due to stress concentration, the crack develops faster in length and its aspect ratio becomes smaller. Since the method used herein enables calculation of stress intensity factors for the deepest point only, a forcing function proposed in [2] was used to account for changes in the crack shape.

Final crack depth was assumed to be equal to a half of the plate thickness. However, by each load cycle a check was made by the program used for the crack growth analysis, whether the current stress intensity factor does not exceed the fracture toughness K_{Ic}, and an appropriate correction was made.

Random variables

The Monte Carlo simulation is often used in the probabilistic fatigue life analysis of real components [1,3-5,15], since it enables using the same deterministic model of fatigue life prediction and it accounts for various types of probability distributions of the selected random variables. The essence of the Monte Carlo method consists in using a computer simulation technique for obtaining a probability distribution of the calculated fatigue life. Multiple calculations are made in order to produce a sufficient number of data. Magnitudes of the source variables are for each calculation randomly sampled from their probability distributions. Hence, probability distributions of those variables make an empirical basis for the method.

There are many factors affecting the scatter in fatigue life of structural components of machines in service. Loading variations due to environment and operating conditions, scatter in the experimentally obtained crack propagation rate, threshold stress intensity factor, residual stresses, shape and dimensions of the initial crack or crack-like defect are the potential sources of that phenomenon. In probabilistic crack growth analyses some of those parameters are usually treated as the source variables. Their probability distributions are determined experimentally or in many cases arbitrarily, whereas different types of distributions are selected sometimes for the same variable. Variables selected as source variables in the present fatigue crack growth analysis are described below along with a short review of assumptions, made in other analyses found in the literature.

Material properties. The material properties, C and m in the Paris' equation are usually considered to be random variables, not necessarily together. Due to reported strong correlation between C and m, the exponent m was assumed as a random variable and values of C were calculated in [5] according to the known correlation relation. However, in probabilistic crack growth analyses, the exponent m is also assumed to be a fixed value, as a result of the regression analysis and only the parameter C is considered a random variable [1,3,4,16,19] following a log-normal distribution. The same assumption was made in the present analysis, and the standard deviation of 0.3 was assumed for the parameter C. The threshold stress intensity factor ΔK_{th} and fracture toughness K_{Ic} are seldom treated as random variables. In the present work a fixed value of ΔK_{th} was assumed of 4 MPa$\sqrt{m}$, while the fracture toughness was modelled as normal distribution with the mean value of 52 MPa$\sqrt{m}$ [8] and the standard deviation of 5 MPa$\sqrt{m}$.

Notch geometry. The overall geometry of a welded joint as well as local geometry of weld notch affect the magnitude of the stress intensity factor. This is due to the stress concentration, caused by changes in the plate thickness. Stress concentration in welded joints depends on the weld profile, described by parameters Θ and ρ (Fig. 1). Engesvik and Lassen [5] have considered those two parameters independent variables, while in the present study, the combined effect of the parameters, i.e. the stress concentration factor was considered. The stress concentration factor for cover plate was calculated using an analytical expression found in [18] and its distribution was modelled by a two-parameter Weibull distribution (Fig. 3). The scatter in stress concentration factor for the component M.2 was significantly larger than that for the component M.1.

Crack shape and dimensions. Assumptions regarding dimensions of cracks or crack-like defects are very important for the fatigue life analysis, since they significantly affect the fatigue crack growth. Various types of probability distributions are assumed for surface cracks e.g. fixed [15], log-normal [19], Weibull [16], exponential [1,4] or shifted exponential [5]. Assumptions made in respect to crack dimensions differ significantly, depending on the object being considered. Hedegard at al. [6] have shown that depending on the welding parameters of the MAG-welding process, maximum depths of flaws found in welded specimens changed from 0.1 to 1.4 mm. The magnitude of 0.5 mm of a mean value of an initial crack depth seems to be reasonably as far as welded components are considered [19]. In this study the crack depth a was assumed to be normally distributed with the mean value of 0.5 mm and standard deviation of 0.1 mm.
The aspect ratio a/c is modelled by uniform [15], normal or log-normal [1, 19] distributions. For the present analysis a fixed shape of the initial crack was assumed of the aspect ratio of $a/c = 1.0$. This assumption was of a small importance, since the initial crack shape was in fact defined by the forcing function used in the analysis and it made the aspect ratio of the crack decrease very rapidly during crack growth.

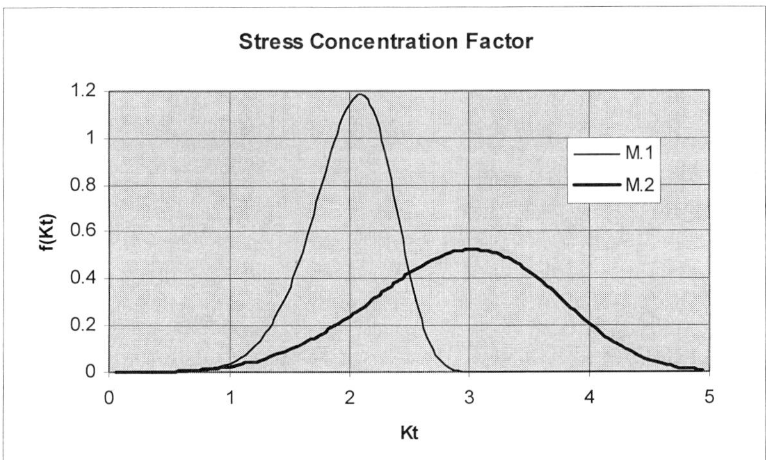

Fig. 3. Weibull distributions of stress concentration factor for welded joints M.1 and M.2

Service loading. Stress history recorded during field operation of a similar excavator was used for the analysis. It was rescaled to the considered point of the structure using a computer program for analysis of working attachment of hydraulic excavators. Service loading varies not only within one duty cycle of a machine, but it also changes for different operation conditions, task realised, operators skills, etc. Since the recorded stress history (Fig. 5) is not very long and it could be considered representative for medium operating conditions only, another scaling factor k was introduced, which accounted for uncertainties in the assumed service loading, e.g. scatter in service loading for different operation conditions. The mean value of the factor k was assumed $k = 1$, and the standard deviation of its normal distribution was 0.05.

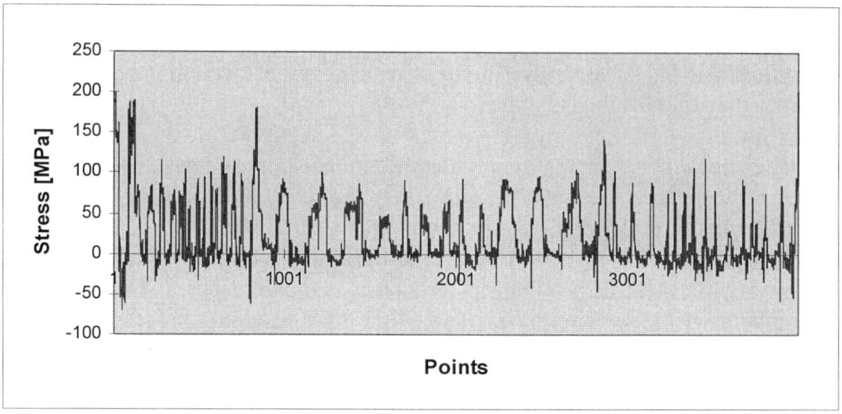

Fig. 4 Stress history used for life predictions

Distributions of all considered random variables were truncated, their end values for each variable are specified in Table 1.

Table 1. Data on source random variables

Variable	Distribution	Range
C	log-normal	4.19e-13 - 2.54e-12
K_{Ic}	normal	35 - 65 [MPa√m]
K_t	Weibull	1.2-3.0 (M.1) and 1.2 - 5.0 (M.2)
a	normal	0.2 - 0.8 [mm]
k	normal	0.85 - 1.15

Probabilistic analysis

A modified computer program FALPR [13] was used for probabilistic fatigue life predictions. The program enables source variables to be specified as fixed or random, whereas one of three types of probability distribution (normal, log-normal, and Weibull) for each random variable can be selected.

Deterministic crack propagation analyses were performed for 500 sets of data, randomly sampled from the probability distributions of the source variables. Each result gave a single point for the probability distribution of the fatigue life. A theoretical distribution was fitted to results of one simulation. The best fit was achieved when two-parameter log-normal or Weibull distributions were used. Fatigue life with 80% probability of no-failure (T_{80}) was estimated from the theoretical distribution, since this level of reliability is acceptable for structural components of earthmoving machines [13].

Engesvik and Lassen [5] have shown, how each of the source variable affects the scatter of the simulated life distribution. However, it is of interest, how the predicted fatigue life changes when not all source variables are considered as random. This may also give one an answer, which variables may be considered as fixed ones. Therefore, the simulations were performed for eight cases, specified in Table 2. In the first case all variables were treated as random, while in the other cases one or more variables were considered fixed with their mean values, except the material properties, which were always kept random. For comparisons, results of the deterministic analyses, performed for fixed values of all source variables (mean values) are shown in Table 2 as well.

The same analyses were carried out with an assumption of a single semi-elliptic surface crack and an edge crack, in order to obtain the upper and lower bounds of the predicted fatigue lives, as it was suggested in [9].

RESULTS AND DISCUSSION

An example of simulation, performed for the case with a fixed value of the load scaling factor *k* and a semi-elliptic crack is shown in Fig. 5. The reliability distribution R(T) was obtained as R(T) = 1 - F(T), where F(T) is the cumulative probability distribution of fatigue life. Results of all simulations performed for the two considered components M.1 and M.2 are summarised in Table 2. All values present the predicted fatigue life T_{80} (80% survival probability) except those for all mean variables, denoted further as mean lives.

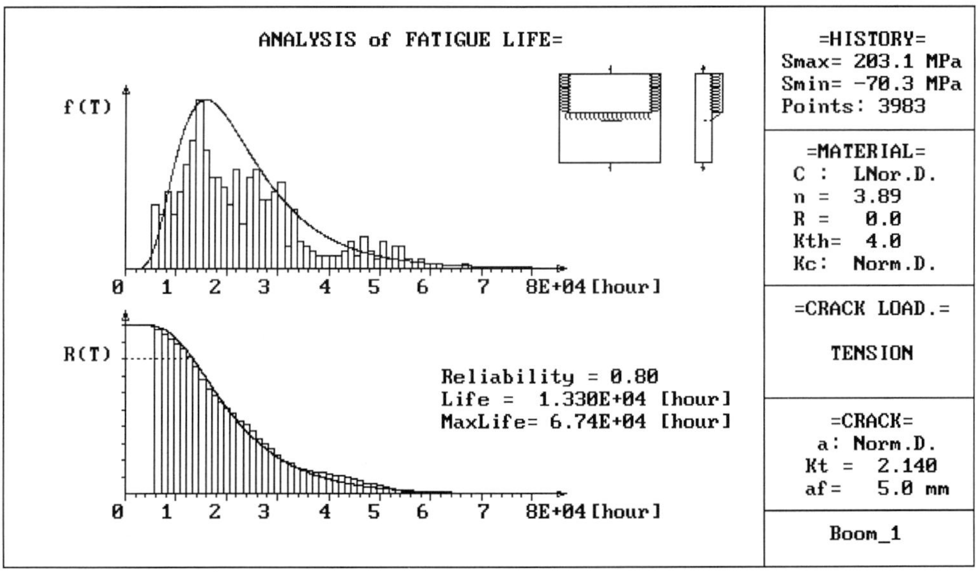

Fig. 5 Example of simulation result (M.1, Kt=2.14)

Results obtained for all considered cases of simulation can be divided in two groups of equal quantities, which gave comparable results of the predicted fatigue lives T_{80}. The first group is built of cases denoted as: *all*, *a*, *k* and a+*k*, and the second of cases *Kt*, *a+Kt*, *k+Kt* and *a+k+Kt*. The second group of simulations is characterised by a significantly smaller scatter due to assuming a fixed value of the stress concentration factor K_t. This can be seen in Fig. 6, where the probability density distributions f(T) for the component denoted M.1 with a semi-elliptic crack are shown, and in Fig. 7, where values of maximal calculated lives T_{max} and predicted T_{80} lives are shown for the same component. It means, that scatter in the weld geometry has the greatest influence on predicted fatigue lives and this coincides with results obtained by other researchers [5].

Table 2. Results of fatigue life prediction [hours]

Case (source variables)	Notation in Fig. 5	M.1		M.2	
		semi-ellipt.	Edge	semi-ellipt.	Edge
All random	all	10570	8414	2746	1943
a=0.5	a	9419	7379	2346	1730
k=1	k	10770	8079	2775	2124
Kt=2.14/3.03	Kt	13300	9163	3197	2401
a=0.5,Kt=2.14/3.03	a+Kt	12110	8705	2968	2268
k=1,Kt=2.14/3.03	k+Kt	12630	8484	3187	2382
a=0.5,k=1	a+k	9518	6670	2467	2007
a=0.5,k=1,Kt=2.14/3.03	a+k+Kt	11070	8420	2982	2192
All mean	mean	19237	10549	5038	2763

The predicted T_{80} lives from the first group of simulations are shorter in comparison to those from the second one. Maximal differences are from ca. 20% to 30% for both components (M.1 and M.2), and both crack shapes. It is worth noting, that similar difference exists between fatigue lives T_{80} predicted for edge cracks and those obtained for semi-elliptic cracks, while the corresponding mean lives are shorter by ca. 45%.

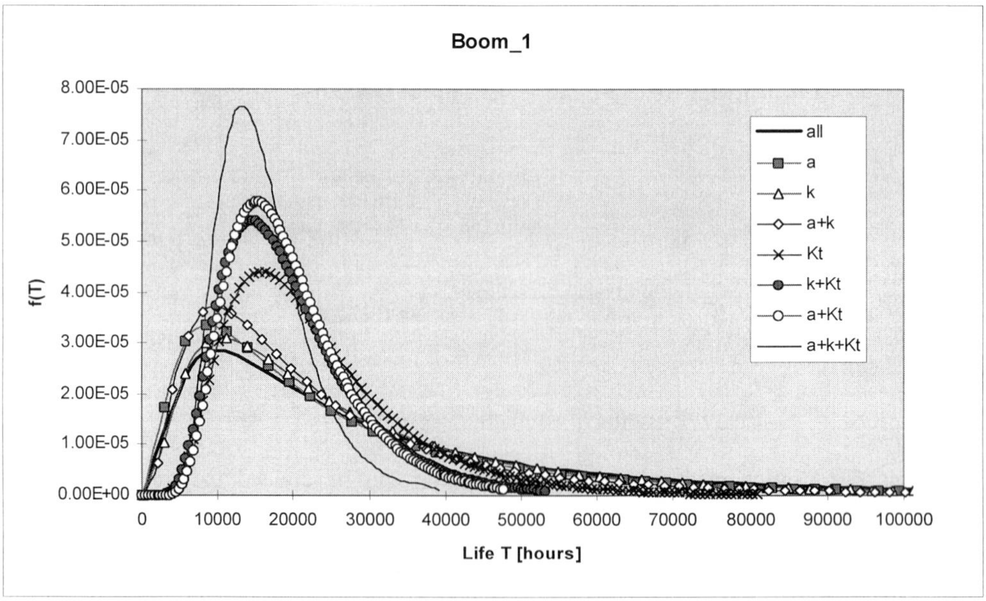

Fig. 6. Fatigue life distributions for the component M.1.

In the presented analysis the scaling factor *k* of service loading had rather a small influence on the result, and it might be neglected. However, this may be valid only for the assumed distribution of the scaling factor. Assuming a fixed value of crack depth resulted generally in decreasing the predicted lives T_{80}. The most optimistic results were obtained for fixed values of the stress concentration factor, Kt = 2.14 and 3.03 for the component M.1 and M2. respectively.

Life distributions for all cases were modelled by a log-normal distribution, however in case *a+k+Kt*, in which only material properties were considered as random, a two-parameter Weibull distribution was also acceptable. Generally, the largest scatter in the simulated fatigue life was obtained when all source variables were considered to be random, while the smallest scatter was obtained when material properties were random only. The weld geometry, defined here by distribution of the stress concentration factor K_t had the greatest impact on the scatter.

The differences in predicted T_{80} lives (Fig. 7) are rather not substantial from the designer's point of view, and it may suggest, that neglecting the scatter in some source variables does not affect the predicted fatigue life seriously. It is not certain if this is a general rule, since it may depend on type and parameters of distributions of considered source variables, and therefore it requires a further research.

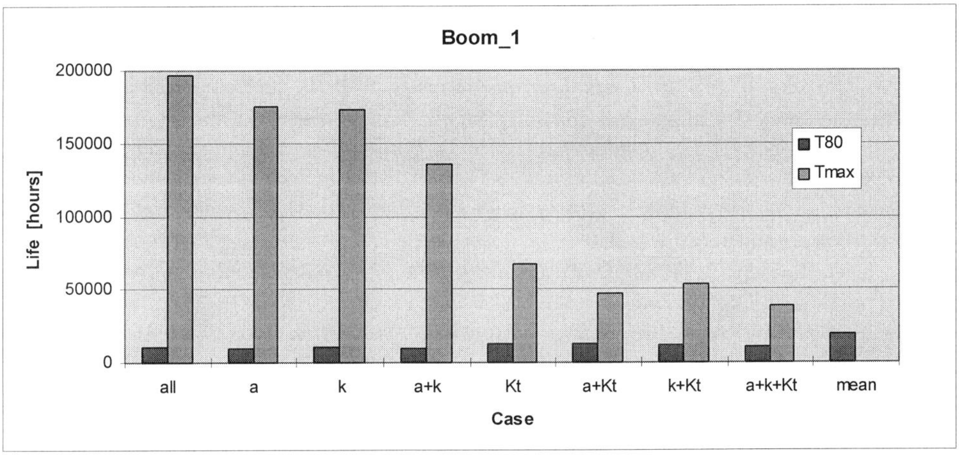

Fig. 7 Comparison of calculated Tmax and T_{80} fatigue lives (semi-elliptic crack)

The effect of manufacturing quality on the predicted fatigue life of components is very distinct. The predicted fatigue lives for the worse component (M.2) are smaller in comparison to those corresponding to the better one (M.1) by factor of ca. 4. The only reason for that is the large difference in magnitudes of the stress concentration factor at the weld toe (Fig. 2). Results obtained in the analysis for the component denoted M.1 as the upper bound (semi-elliptic crack), were acceptable since the required life of 10000 hours was generally fulfilled. However, the lower bound results (edge crack) for the component M.1, as well as all results for the component M.2 were below the required fatigue life. The analysis confirmed the insufficient life of components M.2 observed in service, which meant the manufacturing quality of the component M.2 could not be accepted.

CONCLUSIONS

The presented probabilistic analysis of fatigue life of a structural component with the Monte Carlo simulation was carried out. Fatigue lives T_{80} were predicted for the required level of components reliability of R(T) = 0.8.

The simulations were performed for different cases, assuming various combinations of fixed and random source variables. It was shown, that the scatter of the predicted fatigue life decreased if fewer source variables were considered random. A distinct difference in the scatter was observed when the stress concentration factor, describing the local weld geometry, was considered a fixed variable, in other cases the difference was negligible and thus the predicted fatigue lives were comparable.

For the assumed distributions of source variables, values of predicted T_{80} lives for all combinations of random and fixed source variables do not differ significantly, but this cannot be considered as a general rule. It may be suggested however, that the scatter in weld geometry should not be neglected, since it affects mostly the predicted fatigue life.

Fatigue life predictions confirmed the insufficient resistance against fatigue failure of the component M.2, which resulted from the low level of manufacturing quality.

REFERENCES

1. Akama, M., Ishizuka, H. (1995), *JSME Journal*, Series A, Vol. 38, No. 3, pp. 378-383
2. Almar-Naess, A. (1985), *Fatigue Handbook. Offshore Steel Structures*, Tapir Forlag, Trondheim
3. Brückner-Foit, A., Jäckels, H., Quadfasel, U., (1993), *Fatigue Fract. Engng Mater. Struct.*, Vol. 16, No. 8, pp. 891-908
4. Cohelo da Silva, R.B., Bastian, F.L. (1995), in *Proceedings of the VTT Symposium on Fatigue Design*, G. Marquis and J. Solin (Eds).
5. Engesvik, K.M., Lassen, T. (1988), in *Proceedings of the 7th International Conference on Offshore Structures and Arctic Engineering*
6. Hedegard, J. at al. (1995), *in Proceedings of the VTT Symposium on Fatigue Design*, G. Marquis and J. Solin (Eds).
7. Jakubczak, H., Glinka, G. (1997), in *Proceedings of the 5th International Conference on Biaxial - Multiaxial Fatigue and Fracture*, E. Macha and Z. Mróz (Eds.).
8. Kocañda, S., Szala, J. (1985), *Podstawy obliczen zmeczeniowych*, PWN, Warszawa
9. Lecsek, L.R., Yee, S.B., Lambert, S.B., and Burns, D.J.(1995), *Fatigue Fracture of Engineering Materials and Structures*, Vol. 18, No. 7/8, pp. 821-831.
10. Niu, X., Glinka, G. (1987), *International Journal of Fracture*, No. 35, pp. 3-20.
11. Radaj, D.(1996), *International Journal of Fatigue*, Vol. 18, No3, pp. 153-170.
12. Shen, G., Glinka, G.(1991), *Theoretical and Applied Fracture Mechanics*, Vol. 15, pp. 247-255.
13. Sobczykiewicz, W., Glinka, G., Jakubczak, H.,(1992), *in Proceedings of the VTT Symposium on Fatigue Design*, G. Marquis and J. Solin (Eds).
14. Sobczykiewicz, W., Rzeszot, J. (1992), in *Proceedings of "The V Conf. on Developmen of Earthmoving and Construction Machines"*, Zakopane, Poland.
15. Sutharshana, S.,Creager, M., Ebbler, D., Moore, N. (1992), *ASTM STP 1122*, pp. 234-246.
16. Tryon, R.G, Cruse, T.A., Mahadevan, S. (1996), *Engg. Fracture Mechanics*, Vol.53, No. 5, pp 807-828.
17. Xu, R.X., Thompson, J.C., Topper, T.H.(1995), *Fatigue Fracture of Engineering Materials and Structures,* Vol. 18, No.7/8, pp. 885-895.
18. Yung, J.Y., Lawrence, F.V. (1985), *Report No. 117*, University of Illinois, Urbana-Champaign
19. Zhao, Z., Haldar, A. (1996), *Engineering Fracture Mechanics*, Vol. 53, No. 5, pp. 775-789.

Author Index

Risse (ed.)

The Complete (but unofficial) Guide to the Willem C. Vis International Commercial Arbitration Moot